Kelley Wingate
Math Practice

First Grade

Credits
Content Editor: Jennifer B. Stith
Copy Editor: Karen Seberg

Visit *carsondellosa.com* for correlations to Common Core, state, national, and Canadian provincial standards.

Carson-Dellosa Publishing, LLC
PO Box 35665
Greensboro, NC 27425 USA
carsondellosa.com

ISBN 978-1-4838-0499-6
02-111141151

Table of Contents

Introduction

Competency in basic math skills creates a foundation for the successful use of math principles in the real world. Practicing math skills—in the areas of operations, algebra, place value, fractions, measurement, and geometry—is the best way to improve at them.

This book was developed to help students practice and master basic mathematical concepts. The practice pages can be used first to assess proficiency and later as basic skill practice. The extra practice will help students advance to more challenging math work with confidence. Help students catch up, stay up, and move ahead.

Common Core State Standards (CCSS) Alignment

This book supports standards-based instruction and is aligned to CCSS standards. The standards are listed at the top of each page for easy reference. To help you meet instructional, remediation, and individualization goals, consult the Common Core State Standards alignment chart on page 4.

Leveled Activities

Instructional levels in this book vary. Each area of the book offers multilevel math activities so that learning can progress naturally. There are three levels, signified by one, two, or three dots at the bottom of the page:

- Level I: These activities will offer the most support.
- Level II: Some supportive measures are built in.
- Level III: Students will understand the concepts and be able to work independently.

All children learn at their own rate. Use your own judgment for introducing concepts to children when developmentally appropriate.

Hands-On Learning

Review is an important part of learning. It helps to ensure that skills are not only covered but are internalized. The flashcards at the back of this book will offer endless opportunities for review. Use them for a basic math facts drill, or to play bingo or other fun games.

There is also a certificate template at the back of this book for use as students excel at daily assignments or when they finish a unit.

Common Core State Standards Alignment Chart

Common Core State Standards*		Practice Page(s)
Operations and Algebraic Thinking		
Represent and solve problems involving addition and subtraction.	1.OA.1, 1.OA.2	5–19
Understand and apply properties of operations and the relationship between addition and subtraction.	1.OA.3, 1.OA.4	20–28
Add and subtract within 20.	1.OA.5, 1.OA.6	29–37
Work with addition and subtraction equations.	1.OA.7, 1.OA.8	38–49
Number and Operations in Base Ten		
Extend the counting sequence.	1.NBT.1	50–52
Understand place value.	1.NBT.2, 1.NBT.3	53–64
Use place value understanding and properties of operations to add and subtract.	1.NBT.4–1.NBT.6	65–76
Measurement and Data		
Measure lengths indirectly and by iterating length units.	1.MD.1, 1.MD.2	77-82
Tell and write time.	1.MD.3	83–88
Represent and interpret data.	1.MD.4	89–91
Geometry		
Reason with shapes and their attributes.	1.G.1–1.G.3	92–103

* © Copyright 2010. National Governors Association Center for Best Practices and Council of Chief State School Officers. All rights reserved.

Solving Word Problems: Adding To

For each story, draw a picture and write a number sentence.

Example:
   Brad had 3 balloons.
   His mom got him 2 more.
   Now, he has 5 balloons.

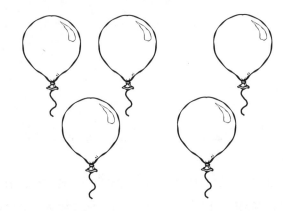

3 + 2 = **5** balloons
_____ = ____ balloons

1. Tia had 4 balloons.
   She blew up 3 more.
   Now, she has 7 balloons.

_____ = _____ balloons

2. Zeb had 2 balloons.
   His sister gave him 4 more.
   Now, he has 6 balloons.

_____ = _____ balloons

3. Rose had 1 balloon.
   Her dad gave her 7 more.
   Now, she has 8 balloons.

_____ = _____ balloons

Solving Word Problems: Adding To

For each story, draw a picture and write a number sentence.

Example: Kelly had 3 table tennis balls. She found 5 more. How many table tennis balls does she have now? ___**3 + 5**___ = __**8**__ table tennis balls	1. Rick had 7 tennis balls. He found 2 more. How many tennis balls does he have now? _____ = _____ tennis balls
2. Leo had 16 baseballs. His friend gave him 3 more. How many baseballs does he have now? _____ = _____ baseballs	3. Jan's team had 5 footballs. The coach bought 6 more at the store. How many footballs does the team have now? _____ = _____ footballs
4. Becky had 18 rubber balls. Her sister gave her 2 more. How many rubber balls does she have now? _____ = _____ rubber balls	5. Mike's team had 2 kickballs. His father gave the team 10 more. How many kickballs does the team have now? _____ = _____ kickballs

Solving Word Problems: Adding To

For each number sentence, write an "adding to" story.

Example:

3 + 2 = 5   I had 3 purple markers. My friend gave me 2 more. Now, I have 5 purple markers.

1. 4 + 2 = 6   _____

_____

_____

2. 7 + 4 = 11   _____

_____

_____

3. 5 + 4 = 9   _____

_____

_____

4. 16 + 3 = 19   _____

_____

_____

Solving Word Problems: Taking From

For each story, draw a picture and write a number sentence.

Example: Brooke had 6 cat books. She gave 2 to her friend. Now, she has 4 books. 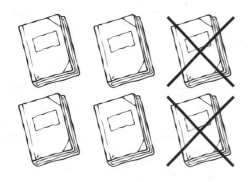 ___**6 – 2**___ = __**4**__ books	1. Tyler had 5 truck books. He lost 1. Now, he has 4 books. _____ = _____ books
2. Lee had 7 joke books. She gave 4 to her brother. Now, she has 3 books. _____ = _____ books	3. Dan had 8 insect books. His dog chewed up 3. Now, he has 5 books. _____ = _____ books

Solving Word Problems: Taking From

For each story, draw a picture and write a number sentence.

Example:
   Ivan had 6 beetles.
   Then, 2 got lost.
   How many beetles were left?

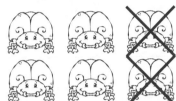

   6 – 2 = **4** beetles

1. Jayla had 5 flies.
   Then, 1 flew away.
   How many flies were left?

   _____ = _____ flies

2. Bill had 17 ants.
   Then, 2 ran away.
   How many ants were left?

   _____ = _____ ants

3. Sally had 8 butterflies.
   Then, 4 fluttered away.
   How many butterflies were left?

   _____ = _____ butterflies

4. Greg had 19 bees.
   Then, 9 went back to the hive.
   How many bees were left?

   _____ = _____ bees

5. Penny had 17 caterpillars.
   Then, 14 crawled away.
   How many caterpillars were left?

   _____ = _____ caterpillars

Solving Word Problems: Taking From

For each number sentence, write a "taking from" story.

Example:
   9 − 2 = 7   I had 9 computer games. I gave my friend 2 games. Now, I have 7 computer games.

1. 16 − 6 = 10 _____

_____

_____

2. 12 − 7 = 5 _____

_____

_____

3. 9 − 3 = 6 _____

_____

_____

4. 18 − 10 = 8 _____

_____

_____

Solving Word Problems: Putting Together/Taking Apart

For each story, draw a picture. Write the answer on the line.

Example:
Patsy has 5 sisters.
She also has 3 brothers.
How many sisters and brothers
does Patsy have in all?

__8__ sisters and brothers

1. Alvin has 2 aunts.
   He also has 4 uncles.
   How many aunts and uncles
   does Alvin have in all?

   _____ aunts and uncles

2. Ben has 2 grandmas.
   He also has 2 grandpas.
   How many grandparents
   does Ben have in all?

   _____ grandparents

3. Lily has 3 boy cousins.
   She also has 4 girl cousins.
   How many cousins does
   Lily have in all?

   _____ cousins

4. Katie has 6 neighbors on the left.
   She also has 3 neighbors on the right.
   How many neighbors does Katie
   have in all?

   _____ neighbors

5. Owen has 1 brother.
   He also has 6 sisters.
   How many brothers and sisters
   does Owen have in all?

   _____ brothers and sisters

Solving Word Problems: Putting Together/Taking Apart

Write a number sentence to solve each problem.

1. Fritz is juggling 8 balls. He has 4 red balls. The rest are green.
   How many balls are green?

2. Mombo has 10 stars on his right shoe. He has 5 stars on his left shoe. How many stars are
   on Mombo's shoes in all?

3. There are 9 polka dots on Mimi's hat. There are 6 polka dots on Sappy's hat. How many
   polka dots are on the hats in all?

4. Bonzo has 12 balloons. He has 9 purple balloons. The rest are pink. How many pink
   balloons does Bonzo have?

Solving Word Problems: Putting Together/Taking Apart

Write a number sentence to solve each problem.

1. Tyler hit 8 balls over the fence. Jack hit 6 balls over the fence. How many balls in all were hit over the fence?	2. There are 6 games in Tara's room. There are 4 games in Jan's room. How many games do the girls have in all?
3. Connor has 9 marbles in his bag. David has 5 marbles. How many marbles do the boys have in all?	4. Two friends took turns running 12 laps around the track. Sierra ran some of the laps. Alex ran 9 of the laps. How many laps did Sierra run?
5. In the game, 11 goals were scored. Kristen scored 7 goals. Kate scored the rest of the goals. How many goals did Kate score?	6. In the first race, Nick swam for 8 minutes. In the next race, he swam for 5 minutes. How many minutes did Nick swim in all?

Solving Word Problems: Comparing

Bill and Becky counted some things they own and made graphs to compare their totals.
Looks at the graphs and answer the questions. Then, write a number sentence to show the
comparison.

1. Board Games

Bill

Becky

Who has more? _____

How many more? _____

_____

2. Hats

Bill

Becky

Who has more? _____

How many more? _____

_____

3. Balls

Bill

Becky

Who has more? _____

How many more? _____

_____

4. Video Games

Bill

Becky

Who has more? _____

How many more? _____

_____

5. Hamsters

Bill

Becky

Who has more? _____

How many more? _____

_____

6. Posters

Bill

Becky

Who has more? _____

How many more? _____

_____

Solving Word Problems: Comparing

Mark and Molly counted some things they own. Read the clues. Shade in the squares to help Mark and Molly complete their graphs. Then, write a number sentence to answer each question.

1. **T-Shirts**

Mark
Molly

Mark has 5. Molly has 2 more.
How many does Molly have? _____

_____

2. **Toy Cars**

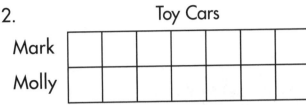

Molly has 6. Mark has 1 more.
How many does Mark have? _____

_____

3. **Building Sets**

Mark
Molly

Mark has 4. Molly has 3 more.
How many does Molly have? _____

_____

4. **Books**

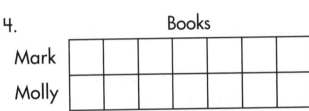

Molly has 3. Mark has 3 more.
How many does Mark have? _____

_____

5. **Stuffed Animals**

Mark
Molly

Mark has 7. Molly has 2 less.
How many does Molly have? _____

_____

6. **Puzzles**

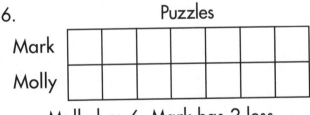

Molly has 6. Mark has 2 less.
How many does Mark have? _____

_____

7. **CDs**

Mark
Molly

Mark has 7. Molly has 3 less.
How many does Molly have? _____

_____

8. **Sweatshirts**

Mark
Molly

Molly has 5. Mark has 1 less.
How many does Mark have? _____

_____

Solving Word Problems: Comparing

Luke and Lily counted the stickers in their collections. Draw a picture and write a number sentence to solve each problem.

1. Luke has 3 fish stickers. Lily has 2 more. How many does Lily have? _____	2. Luke has 6 dog stickers. Lily has 2 less. How many does Lily have? _____
3. Luke has 4 dinosaur stickers. Lily has 3 more. How many does Lily have? _____	4. Luke has 6 space stickers. Lily has 3 less. How many does Lily have? _____
5. Lily has 6 cat stickers. Luke has 1 more. How many does Luke have? _____	6. Lily has 7 elephant stickers. Luke has 2 less. How many does Luke have? _____
7. Lily has 4 ocean stickers. Luke has 3 less. How many does Luke have? _____	8. Lily has 5 bird stickers. Luke has 1 more. How many does Luke have? _____

Solving Word Problems: Three Numbers with Sums to 20

When finding the sum of three addends, look for a fact you know.

$$\begin{array}{r} 6 \\ 2 \\ + 4 \end{array} \Big\rangle 8 \qquad 8 + 4 = 12$$

$$\begin{array}{r} 6 \\ 2 \\ + 4 \end{array} \Big\rangle 6 \qquad 6 + 6 = 12$$

$$\begin{array}{r} 6 \\ 2 \\ + 4 \end{array} \Big\rangle 10 \qquad 10 + 2 = 12$$

Solve each problem. Circle the numbers you would add first.

1. $\begin{array}{r} 3 \\ 5 \\ + 4 \\ \hline \end{array}$

2. $\begin{array}{r} 3 \\ 8 \\ + 6 \\ \hline \end{array}$

3. $\begin{array}{r} 7 \\ 1 \\ + 5 \\ \hline \end{array}$

4. $\begin{array}{r} 6 \\ 4 \\ + 1 \\ \hline \end{array}$

5. $\begin{array}{r} 9 \\ 2 \\ + 3 \\ \hline \end{array}$

6. $\begin{array}{r} 4 \\ 5 \\ + 4 \\ \hline \end{array}$

7. $\begin{array}{r} 2 \\ 1 \\ + 6 \\ \hline \end{array}$

8. $\begin{array}{r} 3 \\ 3 \\ + 5 \\ \hline \end{array}$

9. David ate 1 banana, 8 grapes, and 9 blueberries for breakfast. How many pieces of fruit did he eat in all?

10. Myra drew 5 rabbits, 5 chicks, and 3 sheep in her picture. How many animals did she draw in all?

Solving Word Problems: Three Numbers with Sums to 20

When finding the sum of three addends, look for a fact you know.

$$\begin{matrix} 6 \\ 2 \\ +4 \end{matrix}\Big\rangle\ 8 \qquad 8 + 4 = 12$$

$$\begin{matrix} 6 \\ 2 \\ +4 \end{matrix}\Big\rangle\ 6 \qquad 6 + 6 = 12$$

$$\begin{matrix} 6 \\ 2 \\ +4 \end{matrix}\Big\rangle\ 10 \qquad 10 + 2 = 12$$

Solve each problem. Circle the numbers you would add first.

1. $\begin{matrix} 2 \\ 1 \\ +9 \end{matrix}$

2. $\begin{matrix} 8 \\ 2 \\ +7 \end{matrix}$

3. $\begin{matrix} 1 \\ 7 \\ +3 \end{matrix}$

4. $\begin{matrix} 5 \\ 6 \\ +2 \end{matrix}$

5. Oliver found 2 pennies, 5 dimes, and 8 nickels in his pocket. How many coins did he find in all?

6. Casey counted 7 red birds, 3 blue birds, and 6 yellow birds on her bird feeder. How many birds did she count in all?

7. Erin had 6 marbles, Brad had 4 marbles, and Chelsea had 4 marbles. How many marbles did the children have in all?

8. Pablo swam 7 laps in the morning. He swam 4 more laps after lunch. He swam 2 more laps after dinner. How many laps did Pablo swim in all?

Solving Word Problems: Three Numbers with Sums to 20

Solve each problem. Circle the numbers you would add first.

1. $6 + 4 + 2 =$ ☐

2. $1 + 1 + 1 =$ ☐

3. $5 + 4 + 6 =$ ☐

4. $7 + 3 + 2 =$ ☐

5. $4 + 3 + 7 =$ ☐

6. $4 + 4 + 8 =$ ☐

7. $6 + 2 + 8 =$ ☐

8. $2 + 2 + 4 =$ ☐

9. $9 + 4 + 3 =$ ☐

10. $3 + 4 + 3 =$ ☐

11. Alicia baked 7 chocolate chip cookies, 5 peanut butter cookies, and 6 oatmeal cookies. How many cookies did Alicia bake in all?

12. Gavin bought 3 packs of gum. Each pack has 5 sticks of gum. How many sticks of gum does Gavin have in all?

13. In the pond, 3 turtles, 6 ducks, and 2 swans are swimming. How many animals are swimming in the pond in all?

14. Jill has 7 pink flowers, 6 orange flowers, and 6 red flowers in a vase. How many flowers does Jill have in the vase in all?

Properties of Operations: Commutative Property

Count the dots on each domino. Solve each problem.

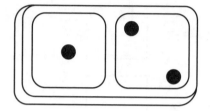

$1 + 2 =$ _____

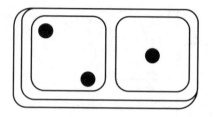

$2 + 1 =$ _____

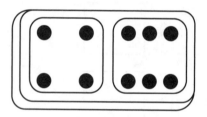

$4 + 6 =$ _____

$6 + 4 =$ _____

$6 + 3 =$ _____

$3 + 6 =$ _____

What do you notice about the two problems in each row?

_____

_____

Properties of Operations: Commutative Property

Connect numbers written in rows to make rainbows. These rainbows will show you addition sentences. Look at the example. Then, make your own rainbows and write the addition sentences that go with them.

Example: The Four Rainbow

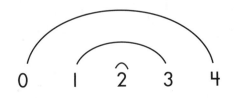

0   1   2   3   4

Number Sentences

$0 + 4 = 4$
$1 + 3 = 4$
$2 + 2 = 4$

1. The Five Rainbow

$= 5$

$= 5$

0   1   2   3   4   5 $= 5$

2. The Seven Rainbow

$= 7$

$= 7$

$= 7$

0   1   2   3   4   5   6   7 $= 7$

3. What do you notice about the two numbers connected by an arch?

_____

_____

1.OA.B.3

Properties of Operations: Commutative Property

Make rainbows for these numbers and write the addition sentences that go with them. The first one has been started for you.

1. The Eight Rainbow

$0 + 8 = 8$

$\qquad = 8$

$\qquad = 8$

$\qquad = 8$

$\qquad = 8$

```
0   1   2   3   4   5   6   7   8
```

2. The Ten Rainbow

$\qquad = 10$

$\qquad = 10$

$\qquad = 10$

$\qquad = 10$

$\qquad = 10$

$\qquad = 10$

```
0   1   2   3   4   5   6   7   8   9   10
```

3. What do you notice about the two numbers connected by an arch?

_____

_____

Properties of Operations: Associative Property

Solve each problem. Circle the two problems in each row that have the same sum.

1. 3 + 1 + 1 =   4 + 1 + 1 =   1 + 3 + 1 =

2. 1 + 2 + 2 =   4 + 6 + 7 =   2 + 1 + 2 =

3. 4 + 6 + 3 =   6 + 4 + 3 =   4 + 0 + 2 =

4. 1 + 1 + 2 =   0 + 4 + 2 =   0 + 2 + 4 =

5. 3 + 2 + 0 =   3 + 1 + 0 =   0 + 2 + 3 =

6. 8 + 1 + 0 =   2 + 1 + 1 =   8 + 0 + 1 =

7. In each row, what do you notice about the numbers in the two problems you circled?

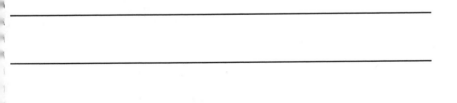

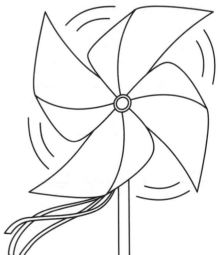

Name _____

Properties of Operations: Associative Property

Solve each problem. Circle the two problems in each row that have the same sum.

1. $2 + 0 + 3 =$ $5 + 3 + 1 =$ $3 + 2 + 0 =$

2. $6 + 3 + 1 =$ $6 + 1 + 3 =$ $4 + 0 + 2 =$

3. $3 + 0 + 4 =$ $3 + 3 + 3 =$ $4 + 3 + 0 =$

4. $8 + 0 + 1 =$ $8 + 1 + 0 =$ $2 + 3 + 2 =$

5. $5 + 1 + 1 =$ $3 + 6 + 1 =$ $1 + 5 + 1 =$

6. $7 + 2 + 1 =$ $1 + 2 + 7 =$ $6 + 1 + 0 =$

7. In each row, what do you notice about the numbers in the two problems you circled?

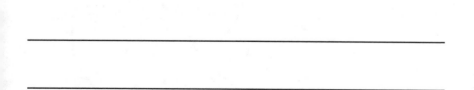

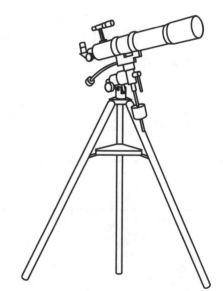

_____

_____

_____

Properties of Operations: Associative Property

Solve the problems in each row. Draw lines to connect the problems that have the same sum. The first one has been done for you.

1.
$$\begin{array}{r} 9 \\ 0 \\ +\ 4 \\ \hline \mathbf{13} \end{array}$$
$$\begin{array}{r} 10 \\ 1 \\ +\ 5 \\ \hline \end{array}$$
$$\begin{array}{r} 2 \\ 3 \\ +\ 6 \\ \hline \end{array}$$
$$\begin{array}{r} 3 \\ 4 \\ +\ 7 \\ \hline \end{array}$$
$$\begin{array}{r} 2 \\ 8 \\ +\ 5 \\ \hline \end{array}$$
$$\begin{array}{r} 6 \\ 1 \\ +\ 5 \\ \hline \end{array}$$

2.
$$\begin{array}{r} 10 \\ 5 \\ +\ 1 \\ \hline \end{array}$$
$$\begin{array}{r} 2 \\ 5 \\ +\ 8 \\ \hline \end{array}$$
$$\begin{array}{r} 4 \\ 0 \\ +\ 9 \\ \hline \mathbf{13} \end{array}$$
$$\begin{array}{r} 2 \\ 6 \\ +\ 3 \\ \hline \end{array}$$
$$\begin{array}{r} 4 \\ 3 \\ +\ 7 \\ \hline \end{array}$$
$$\begin{array}{r} 5 \\ 1 \\ +\ 6 \\ \hline \end{array}$$

3.
$$\begin{array}{r} 5 \\ 7 \\ +\ 4 \\ \hline \end{array}$$
$$\begin{array}{r} 1 \\ 7 \\ +\ 6 \\ \hline \end{array}$$
$$\begin{array}{r} 6 \\ 2 \\ +\ 4 \\ \hline \end{array}$$
$$\begin{array}{r} 9 \\ 5 \\ +\ 1 \\ \hline \end{array}$$
$$\begin{array}{r} 0 \\ 12 \\ +\ 6 \\ \hline \end{array}$$
$$\begin{array}{r} 3 \\ 6 \\ +\ 2 \\ \hline \end{array}$$

4.
$$\begin{array}{r} 1 \\ 5 \\ +\ 9 \\ \hline \end{array}$$
$$\begin{array}{r} 7 \\ 1 \\ +\ 6 \\ \hline \end{array}$$
$$\begin{array}{r} 5 \\ 4 \\ +\ 7 \\ \hline \end{array}$$
$$\begin{array}{r} 3 \\ 2 \\ +\ 6 \\ \hline \end{array}$$
$$\begin{array}{r} 12 \\ 0 \\ +\ 6 \\ \hline \end{array}$$
$$\begin{array}{r} 4 \\ 6 \\ +\ 2 \\ \hline \end{array}$$

5. What do you notice about the numbers in each pair of problems you connected with lines?

_____

_____

Understanding Subtraction as an Unknown Addend Problem

Draw a picture to solve each problem. Then, write an addition sentence and a subtraction sentence for each problem.

Ex.

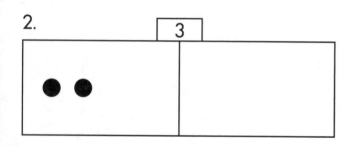

$$\underline{\ \ 6\ \ } + \underline{\ \ ?\ \ } = \underline{\ \ 8\ \ }$$
$$\underline{\ \ 8\ \ } - \underline{\ \ 6\ \ } = \underline{\ \ 2\ \ }$$

1.

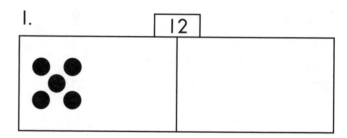

_____ + _____ = _____

_____ − _____ = _____

2.

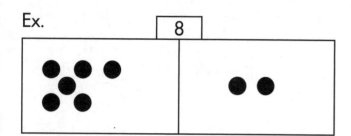

_____ + _____ = _____

_____ − _____ = _____

3.

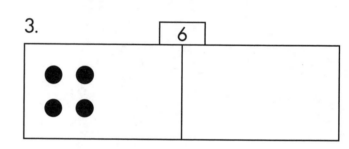

_____ + _____ = _____

_____ − _____ = _____

4.

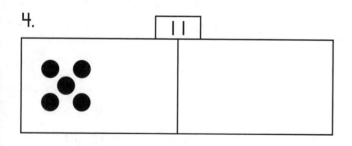

_____ + _____ = _____

_____ − _____ = _____

5.

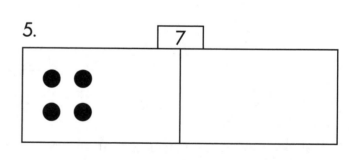

_____ + _____ = _____

_____ − _____ = _____

Understanding Subtraction as an Unknown Addend Problem

The Party Stuff Store counted up all of its party stuff and made some graphs to compare the totals. Find the difference for each graph. Write an addition sentence and a subtraction sentence about each difference. Circle the difference in each sentence.

Example:

red
balloons
yellow
balloons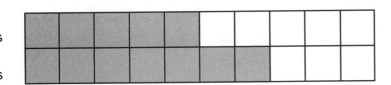

Difference is ___2___.

___5___ + (**2**) = ___7___

___7___ − (**2**) = ___5___

blue
streamers
purple
streamers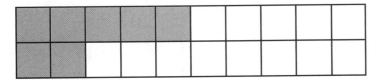

1. Difference is _____.

_____ + _____ = _____

_____ − _____ = _____

plastic
goody bags
paper
goody bags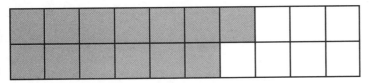

2. Difference is _____.

_____ + _____ = _____

_____ − _____ = _____

paper
confetti
foil
confetti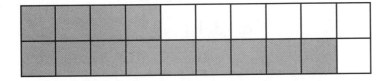

3. Difference is _____.

_____ + _____ = _____

_____ − _____ = _____

tiaras
cone
party hats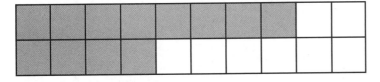

4. Difference is _____.

_____ + _____ = _____

_____ − _____ = _____

paper
plates
plastic
cups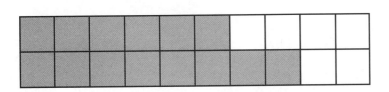

5. Difference is _____.

_____ + _____ = _____

_____ − _____ = _____

Understanding Subtraction as an Unknown Addend Problem

Write a subtraction sentence to solve for the missing number in each number sentence.

1. $6 + \boxed{?} = 12$

_____

2. $7 + \boxed{?} = 12$

_____

3. $8 + \boxed{?} = 11$

_____

4. $7 + \boxed{?} = 14$

_____

5. $5 + \boxed{?} = 13$

_____

6. $3 + \boxed{?} = 13$

_____

Relating Addition and Subtraction

Look at each number in the middle column. Write the number that is 3 less and the number that is 3 more for each. The first one has been done for you.

3 less   3 more   3 less   3 more

1. __1__ 4 __7__   2. ____ 3 ____

3. ____ 6 ____   4. ____ 7 ____

5. ____ 5 ____   6. ____ 10 ____

7. ____ 8 ____   8. ____ 13 ____

9. ____ 11 ____   10. ____ 22 ____

11. ____ 15 ____   12. ____ 35 ____

Relating Addition and Subtraction

Look at each number in the middle column. Write the number that is 5 less and the number that is 5 more for each.

5 less   5 more   5 less   5 more

1. _____ 6 _____   2. _____ 8 _____

3. _____ 7 _____   4. _____ 9 _____

5. _____ 11 _____   6. _____ 15 _____

7. _____ 12 _____   8. _____ 14 _____

9. _____ 19 _____   10. _____ 22 _____

11. _____ 25 _____   12. _____ 28 _____

Relating Addition and Subtraction

You can use a number line to help you add.

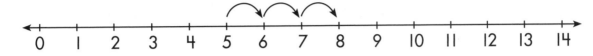

$5 + 3 =$ _____   Start at 5. Count on 3 more—6, 7, 8. $5 + 3 = 8$

Use a number line to help you find the sums.

1. $4 + 6 =$ _____   2. $6 + 8 =$ _____

3. $8 + 4 =$ _____   4. $5 + 5 =$ _____

5. $7 + 7 =$ _____   6 $9 + 3 =$ _____

7. $5 + 4 =$ _____   8. $7 + 6 =$ _____

Use a number line to subtract. Start at the highest number and then count back. The number you land on is your answer.

Example: $14 - 8 = 6$

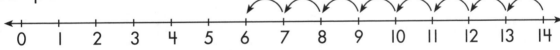

Use a number line to help you find the difference.

9.   12
    − 8
    ____

10.   11
    − 7
    ____

11.   13
    − 4
    ____

12.   14
    − 7
    ____

13.   12
    − 7
    ____

14.   14
    − 8
    ____

15.   11
    − 5
    ____

16.   12
    − 5
    ____

Addition to 20

Solve each problem.

1. $2 + 3 =$

2. $1 + 5 =$

3. $4 + 2 =$

4. $0 + 4 =$

5. $1 + 2 =$

6. $1 + 1 =$

7. $2 + 2 =$

8. $3 + 2 =$

9. $4 + 2 =$

10. $5 + 1 =$

11. $3 + 0 =$

12. $3 + 3 =$

Addition to 20

Solve each problem.

1. $\begin{array}{r} 10 \\ +\ 1 \\ \hline \end{array}$

2. $\begin{array}{r} 4 \\ +\ 7 \\ \hline \end{array}$

3. $\begin{array}{r} 3 \\ +\ 6 \\ \hline \end{array}$

4. $\begin{array}{r} 8 \\ +\ 3 \\ \hline \end{array}$

5. $\begin{array}{r} 7 \\ +\ 5 \\ \hline \end{array}$

6. $\begin{array}{r} 5 \\ +\ 7 \\ \hline \end{array}$

7. $\begin{array}{r} 10 \\ +\ 2 \\ \hline \end{array}$

8. $\begin{array}{r} 3 \\ +\ 7 \\ \hline \end{array}$

9. $\begin{array}{r} 4 \\ +\ 8 \\ \hline \end{array}$

10. $\begin{array}{r} 9 \\ +\ 2 \\ \hline \end{array}$

11. $\begin{array}{r} 1 \\ +\ 9 \\ \hline \end{array}$

12. $\begin{array}{r} 5 \\ +\ 3 \\ \hline \end{array}$

13. $\begin{array}{r} 6 \\ +\ 6 \\ \hline \end{array}$

14. $\begin{array}{r} 7 \\ +\ 3 \\ \hline \end{array}$

15. $\begin{array}{r} 7 \\ +\ 4 \\ \hline \end{array}$

1.OA.C.6

Addition to 20

Solve each problem.

1. $\begin{array}{r} 1 \\ + 8 \\ \hline \end{array}$

2. $\begin{array}{r} 15 \\ + 3 \\ \hline \end{array}$

3. $\begin{array}{r} 10 \\ + 8 \\ \hline \end{array}$

4. $\begin{array}{r} 12 \\ + 8 \\ \hline \end{array}$

5. $\begin{array}{r} 8 \\ + 9 \\ \hline \end{array}$

6. $\begin{array}{r} 9 \\ + 3 \\ \hline \end{array}$

7. $\begin{array}{r} 11 \\ + 2 \\ \hline \end{array}$

8. $\begin{array}{r} 11 \\ + 7 \\ \hline \end{array}$

9. $\begin{array}{r} 12 \\ + 5 \\ \hline \end{array}$

10. $\begin{array}{r} 8 \\ + 2 \\ \hline \end{array}$

11. $\begin{array}{r} 8 \\ + 5 \\ \hline \end{array}$

12. $\begin{array}{r} 9 \\ + 9 \\ \hline \end{array}$

13. $\begin{array}{r} 17 \\ + 3 \\ \hline \end{array}$

14. $\begin{array}{r} 9 \\ + 7 \\ \hline \end{array}$

15. $\begin{array}{r} 13 \\ + 5 \\ \hline \end{array}$

Subtraction from 20

Solve each problem.

1.　　2
　　　− 0
　　　‾‾‾

2.　　5
　　　− 5
　　　‾‾‾

3.　　3
　　　− 1
　　　‾‾‾

4.　　3
　　　− 3
　　　‾‾‾

5.　　3
　　　− 2
　　　‾‾‾

6.　　5
　　　− 3
　　　‾‾‾

7.　　6
　　　− 0
　　　‾‾‾

8.　　2
　　　− 2
　　　‾‾‾

9.　　4
　　　− 0
　　　‾‾‾

10.　　6
　　　− 3
　　　‾‾‾

11.　　1
　　　− 1
　　　‾‾‾

12.　　5
　　　− 4
　　　‾‾‾

13.　　1
　　　− 0
　　　‾‾‾

14.　　4
　　　− 1
　　　‾‾‾

15.　　3
　　　− 3
　　　‾‾‾

Subtraction from 20

Solve each problem.

1. $\begin{array}{r} 12 \\ -8 \\ \hline \end{array}$

2. $\begin{array}{r} 9 \\ -2 \\ \hline \end{array}$

3. $\begin{array}{r} 10 \\ -9 \\ \hline \end{array}$

4. $\begin{array}{r} 11 \\ -5 \\ \hline \end{array}$

5. $\begin{array}{r} 9 \\ -5 \\ \hline \end{array}$

6. $\begin{array}{r} 6 \\ -4 \\ \hline \end{array}$

7. $\begin{array}{r} 10 \\ -5 \\ \hline \end{array}$

8. $\begin{array}{r} 11 \\ -4 \\ \hline \end{array}$

9. $\begin{array}{r} 8 \\ -1 \\ \hline \end{array}$

10. $\begin{array}{r} 12 \\ -3 \\ \hline \end{array}$

11. $\begin{array}{r} 10 \\ -4 \\ \hline \end{array}$

12. $\begin{array}{r} 8 \\ -3 \\ \hline \end{array}$

13. $\begin{array}{r} 12 \\ -4 \\ \hline \end{array}$

14. $\begin{array}{r} 7 \\ -0 \\ \hline \end{array}$

15. $\begin{array}{r} 9 \\ -6 \\ \hline \end{array}$

Subtraction from 20

Solve each problem.

1. $\begin{array}{r} 10 \\ -\ 6 \\ \hline \end{array}$

2. $\begin{array}{r} 12 \\ -\ 7 \\ \hline \end{array}$

3. $\begin{array}{r} 13 \\ -\ 4 \\ \hline \end{array}$

4. $\begin{array}{r} 14 \\ -\ 4 \\ \hline \end{array}$

5. $\begin{array}{r} 20 \\ -\ 8 \\ \hline \end{array}$

6. $\begin{array}{r} 12 \\ -\ 5 \\ \hline \end{array}$

7. $\begin{array}{r} 16 \\ -\ 5 \\ \hline \end{array}$

8. $\begin{array}{r} 18 \\ -\ 9 \\ \hline \end{array}$

9. $\begin{array}{r} 10 \\ -\ 2 \\ \hline \end{array}$

10. $\begin{array}{r} 17 \\ -\ 3 \\ \hline \end{array}$

11. $\begin{array}{r} 11 \\ -\ 3 \\ \hline \end{array}$

12. $\begin{array}{r} 20 \\ -\ 7 \\ \hline \end{array}$

13. $\begin{array}{r} 15 \\ -\ 7 \\ \hline \end{array}$

14. $\begin{array}{r} 19 \\ -\ 9 \\ \hline \end{array}$

15. $\begin{array}{r} 18 \\ -\ 3 \\ \hline \end{array}$

The Equal Sign

Color the two bugs in each box with equal answers. Then, write a number sentence to show they are equal.

1.

2 + 2   7 − 3

4 + 1   5 − 2

_____ = _____

2.
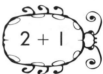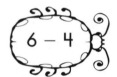

8 − 5   5 + 3

2 + 1   6 − 4

_____ = _____

3.
1 + 1   7 − 5

6 − 3   3 + 4

_____ = _____

4.

7 − 2   4 + 4

5 + 3   8 − 1

_____ = _____

5.
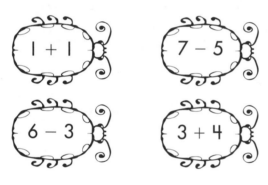

5 + 2   8 − 6

6 − 3

2 − 0   2 + 2

_____ = _____

6.
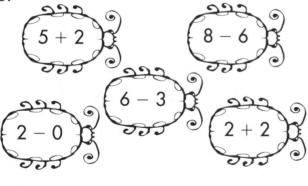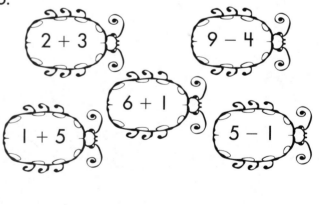

2 + 3   9 − 4

6 + 1

1 + 5   5 − 1

_____ = _____

The Equal Sign

Cut out all of the boxes on the right side of the page. Then, glue the boxes in the spaces to make true number sentences.

1. ☐ = ☐

2. ☐ = ☐

3. ☐ = ☐

4. ☐ = ☐

5. ☐ = ☐

6. ☐ = ☐

$6 + 1$	$4 + 4$	$7 - 2$
$2 + 2$	$5 - 1$	$9 - 2$
$3 + 2$	$6 - 3$	$7 - 1$
$2 + 1$	$3 + 3$	$10 - 2$

The Equal Sign

Find two sets of numbers around the square that are equal. Then, connect their dots with a straight line. Write the number pairs together in a complete number sentence on a line below. Repeat until all of the equal pairs of numbers have been found.

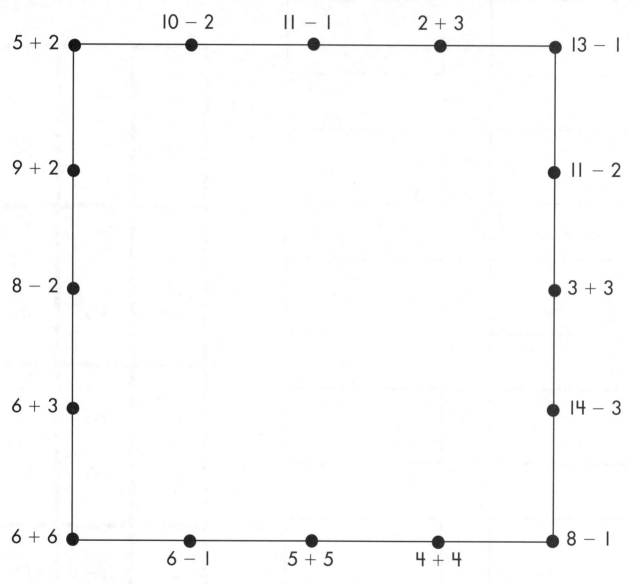

_____   _____

_____   _____

_____   _____

_____   _____

True Number Sentences

Look at the numbers on the ends of each balance. Should they be balanced? Circle yes or no.

Example:

(yes) no   because 3 + 3 = 6 and 2 + 4 = 6

1.

yes   no

2.

yes   no

3.

yes   no

4.

yes   no

5.

yes   no

6.

yes   no

7.

yes   no

8.

yes   no

9.

yes   no

10.

yes   no

11.

yes   no

12.

yes   no

True Number Sentences

Circle all of the number sentences below that are true. Then, answer the questions.

7 = 8	7 = 7
4 + 3 = 7	7 = 4 + 3
4 + 4 = 7	7 = 4 + 4
7 + 4 = 3	7 = 3 + 4
4 + 3 = 4 + 3	4 + 3 = 3 + 4
4 + 3 = 4 + 4	4 + 3 = 1

1. Why is 4 + 3 = 3 + 4 true?

_____

_____

2. Why is 4 + 3 = 4 + 4 not true?

_____

_____

Name _____

True Number Sentences

Write numbers in the blanks to make these number sentences true.

1. _____ = 3

2. _____ = 4

3. _____ = 5

4. _____ = 9

5. 7 = _____

6. 17 = _____

7. 10 = _____

8. 88 = _____

9. _____ + _____ = 6

10. 8 = _____ + _____

11. _____ + _____ = 4

12. 5 = _____ + _____

13. _____ + 1 = _____

14. _____ = 6 + _____

15. _____ + 3 = _____

16. _____ = 2 + _____

17. 5 + _____ = _____

18. _____ = _____ + 3

19. 4 + _____ = _____

20. _____ = _____ + 1

21. _____ + _____ + _____ = 7

22. 9 = _____ + _____ + _____

Name _____

Addition with an Unknown Number

Draw the missing pictures. Finish the number sentences.

1.
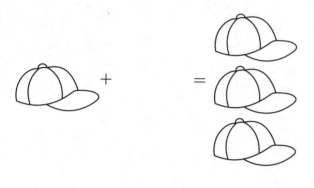

1 + _____ = 3

2.

3 + _____ = 5

3.
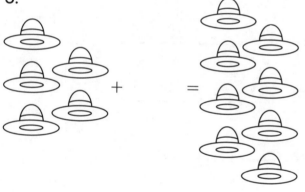

5 + _____ = 8

4.

3 + _____ = 6

5.
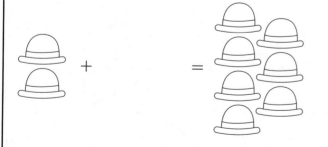

2 + _____ = 7

6.
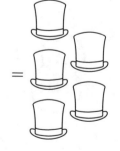

4 + _____ = 5

Addition with an Unknown Number

Draw a picture to solve each problem. Write the answer in the blank.

Example: 4 + __**2**__ = 6

○○   ○
○○   ○

1. 5 = 2 + _____

2. 3 + _____ = 9

3. 8 = 6 + _____

4. _____ + 4 = 4

5. 7 = _____ + 1

6. _____ + 2 = 10

7. 8 = _____ + 4

8. 6 + 3 = _____

9. _____ = 5 + 5

10. 11 + _____ = 15

Addition with an Unknown Number

The Math Whiz is great at math, but she's not perfect. See if you can find her mistakes. Draw a picture to solve the problem. Write the correct answer in the blank to complete the number sentence. Tell whether she is right.

1. $6 + \underline{} = 10$ The Math Whiz says 4. Is she right? _____	2. $7 = 6 + \underline{}$ The Math Whiz says 1. Is she right? _____
3. $7 - \underline{} = 2$ The Math Whiz says 5. Is she right? _____	4. $3 = 4 - \underline{}$ The Math Whiz says 7. Is she right? _____
5. $\underline{} + 4 = 9$ The Math Whiz says 5. Is she right? _____	6. $10 = \underline{} + 8$ The Math Whiz says 18. Is she right? _____
7. $\underline{} - 4 = 2$ The Math Whiz says 2. Is she right? _____	8. $1 = \underline{} - 5$ The Math Whiz says 6. Is she right? _____

Subtraction with an Unknown Number

Draw a picture to solve each problem. Write the answer in the blank.

Example: $9 - 6 =$ _____

⊗ ⊗ ⊗
⊗ ⊗ ⊗
○ ○ ○

1. _____ $= 6 - 2$

2. $7 - 4 =$ _____

3. _____ $= 9 - 7$

4. $5 -$ _____ $= 2$

5. $1 = 4 -$ _____

6. $6 -$ _____ $= 6$

7. $7 = 8 -$ _____

8. _____ $- 3 = 3$

9. $1 =$ _____ $- 4$

10. $12 -$ _____ $= 9$

Subtraction with an Unknown Number

Pluswoman and Minusman have both given an answer to each problem below. Circle the answer that is correct.

1. _____ − 3 = 2 Pluswoman says 4. Minusman says 5.	2. 4 = _____ − 1 Pluswoman says 5. Minusman says 3.
3. _____ − 2 = 5 Pluswoman says 7. Minusman says 3.	4. 2 = _____ − 4 Pluswoman says 2. Minusman says 6.
5. _____ − 1 = 7 Pluswoman says 6. Minusman says 8.	6. 6 = _____ − 2 Pluswoman says 8. Minusman says 4.
7. _____ − 3 = 5 Pluswoman says 2. Minusman says 8.	8. 4 = _____ − 3 Pluswoman says 1. Minusman says 7.

Name _____

Subtraction with an Unknown Number

Find the fish with the correct answer and shade it. Write the answer in the blank to complete each number sentence.

1. $17 - \underline{\hspace{1cm}} = 5$

2. $18 - \underline{\hspace{1cm}} = 15$

3. $8 = \underline{\hspace{1cm}} - 3$

4. $6 - 4 = \underline{\hspace{1cm}}$

5. $\underline{\hspace{1cm}} - 1 = 9$

6. $5 = \underline{\hspace{1cm}} - 2$

7. $\underline{\hspace{1cm}} - 2 = 9$

8. $6 = \underline{\hspace{1cm}} - 6$

Counting to 120

Write the correct number in each box.

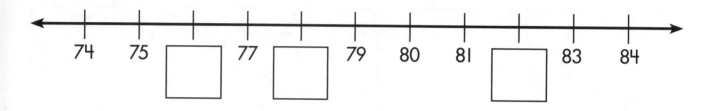

74 75 ☐ 77 ☐ 79 80 81 ☐ 83 84

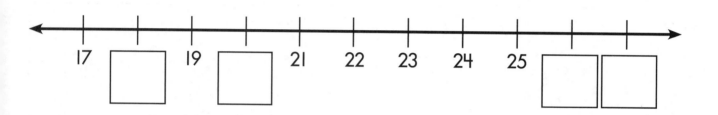

17 ☐ 19 ☐ 21 22 23 24 25 ☐ ☐

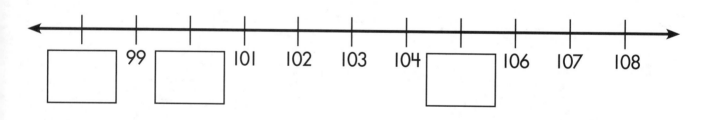

☐ 99 ☐ 101 102 103 104 ☐ 106 107 108

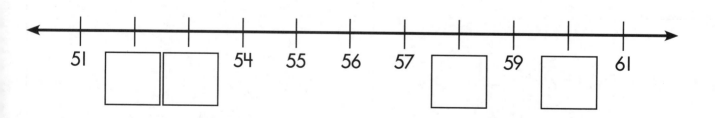

51 ☐ ☐ 54 55 56 57 ☐ 59 ☐ 61

Counting to 120

Write each missing number.

1	2				6		8	9	
11				16					
			25		27	28			30
	32		35					39	
41		43		46	47				50
	52		54	55			58		
61			64			67	68		
		73		75				79	80
	82					87			
		93			96				100
101	102	103		105		107	108		110
111			114				118	119	

Counting to 120

Write the number that comes one before.

1. [ ] [5]

2. [ ] [89]

3. [ ] [23]

4. [ ] [117]

Write the number that comes between.

5. [11] [ ] [13]

6. [35] [ ] [37]

7. [62] [ ] [64]

8. [98] [ ] [100]

Write the number that comes one after.

9. [79] [ ]

10. [92] [ ]

11. [50] [ ]

12. [114] [ ]

13. [41] [ ]

14. [103] [ ]

Tens and Ones

Circle the correct number of blocks to match the number on each gift.

1.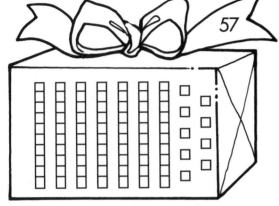

2.

3.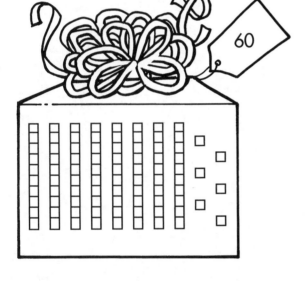

4.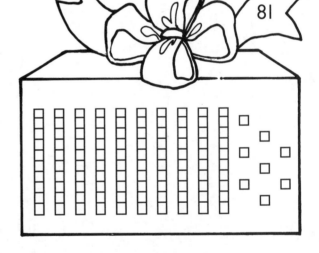

5.

6.

Tens and Ones

Draw lines to match the equal numbers.

49 •

52 •

• 1 ten
  2 ones

• 4 tens
  9 ones

12 •

• 5 tens
  2 ones

55 •

• 5 tens
  5 ones

35 •

• 3 tens
  5 ones

Tens and Ones

Write the number for each group of blocks. Then, use the code to answer the riddle.

Why does a math teacher comb her hair?

To get out the 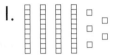 $\frac{\quad}{37}$ $\frac{\quad}{64}$ $\frac{\quad}{73}$ $\frac{\quad}{45}$ $\frac{\quad}{20}$ $\frac{\quad}{35}$ $\frac{\quad}{28}$ $\frac{\quad}{54}$ $\frac{\quad}{64}$ $\frac{\quad}{15}$!

1. _____ → **T**

2. _____ → **G**

3. _____ → **E**

4. _____ → **R**

5. _____ → **S**

6. _____ → **L**

7. _____ → **C**

8. _____ → **N**

9. _____ → **A**

Place Value

Write how many tens and ones are in each number.

52 = _____ tens _____ ones

47 = _____ tens _____ ones

21 = _____ tens _____ ones

36 = _____ tens _____ ones

63 = _____ tens _____ ones

97 = _____ tens _____ ones

18 = _____ tens _____ ones

Place Value

Write each number.

1. 5 tens   4 ones = _____

2. 2 tens   7 ones = _____

3. 8 tens   9 ones = _____

4. 7 tens   5 ones = _____

5. 1 ten   6 ones = _____

6. 4 tens   3 ones = _____

7. 6 tens   0 ones = _____

8. 9 tens   1 one = _____

9. 0 tens   8 ones = _____

10. 3 tens   2 ones = _____

Write the number of tens and ones for each number.

11. 71 = _____ tens   _____ one

12. 58 = _____ tens   _____ ones

13. 5 = _____ tens   _____ ones

14. 40 = _____ tens   _____ ones

1.NBT.B.2

© Carson-Dellosa • CD-104626

Place Value

Write each number in three different ways.

1. 74 = _____ tens _____ ones

   74 = _____ tens _____ ones

   74 = _____ tens _____ ones

2. 88 = _____ tens _____ ones

   88 = _____ tens _____ ones

   88 = _____ tens _____ ones

3. 46 = _____ tens _____ ones

   46 = _____ tens _____ ones

   46 = _____ tens _____ ones

4. 63 = _____ tens _____ ones

   63 = _____ tens _____ ones

   63 = _____ tens _____ ones

Understanding Two-Digit Numbers

Use the code to color the jar of jelly beans.

Color the jelly beans with 0 ones orange.
Color the jelly beans with 2 tens green.
Color the jelly beans with 3 tens red.
Color the jelly beans with 7 tens yellow.
Color the jelly beans with 9 ones purple.
Color the jelly beans with 6 ones blue.

Understanding Two-Digit Numbers

Write the numbers on the left in the correct sets.

1.

6	5	8	2
3	4	7	10
15	24	37	41
1	11	25	36

Has Tens and Ones

Has Only Ones

2.

13	16	12	2
70	50	1	18
0	17	5	3
40	90	15	20

Has Exactly 1 Ten

Has 0 Ones

3.

21	52	42	27
5	12	38	3
2	24	62	32
19	20	72	16

Has Exactly 2 Tens

Has Exactly 2 Ones

Understanding Two-Digit Numbers

Some numbers have been sorted into sets. Cut out and glue each set name under the correct set.

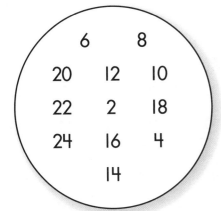

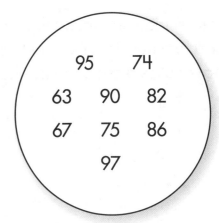

1. _____

2. _____

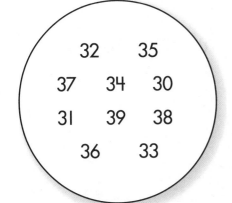

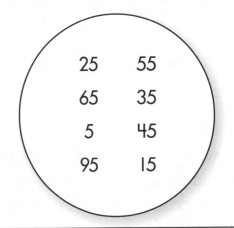

3. _____

4. _____

less than 3 tens	5 in the ones place
more than 8 ones	more than 5 tens
3 in the tens place	less than 2 ones
5 in the tens place	3 in the ones place

Comparing Numbers

Follow the directions for each set of numbers.

1. Draw a triangle around the number that is equal to 7.
   Circle the numbers that are less than 7.
   Draw boxes around the numbers that are greater than 7.

   0   1   2   3   4   5   6   7   8   9   10   11   12   13   14

2. Draw a triangle around the number that is equal to 25.
   Circle the numbers that are less than 25.
   Draw boxes around the numbers that are greater than 25.

   19   20   21   22   23   24   25   26   27   28   29   30   31

3. Draw a triangle around the number that is equal to 30.
   Circle the numbers that are less than 30.
   Draw boxes around the numbers that are greater than 30.

   27   28   29   30   31   32   33   34   35   36   37   38   39

4. Draw a triangle around the number that is equal to 46.
   Circle the numbers that are less than 46.
   Draw boxes around the numbers that are greater than 46.

   38   39   40   41   42   43   44   45   46   47   48   49   50

Comparing Numbers

Write the number in each box that answers the riddle.

1.

I am less than 10.
I am greater than 5.
I am not equal to 8.

What number am I? _____

2.

I am less than 15.
I am greater than 10.
I am not equal to 12.

What number am I? _____

3.

I am less than 6.
I am greater than 3.
I am not equal to 4.

What number am I? _____

4.

I am less than 20.
I am greater than 13.
I am not equal to 15.

What number am I? _____

5.

I am less than 30.
I am greater than 24.
I am not equal to 26.

What number am I? _____

6.

I am less than 40.
I am greater than 30.
I am not equal to 35.

What number am I? _____

Comparing Numbers

Write the numbers in order from least to greatest on the mailboxes.

1.   53   27   19
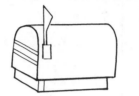 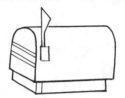 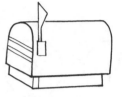

2.   48   31   64
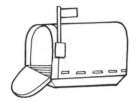 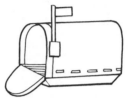 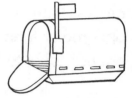

3.   12   36   25
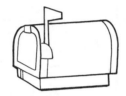 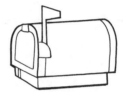 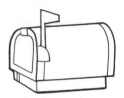

4.   83   74   67
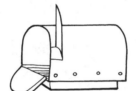 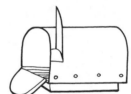 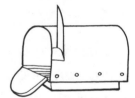

Use <, >, or = to compare the numbers. Use the numbers above to help you.

5. 19 ( ) 27

6. 48 ( ) 31

7. 12 ( ) 36

8. 83 ( ) 83

Adding a Two-Digit Number and a One-Digit Number

Regrouping means to take 10 ones to make another ten. This is how it works.

I have 25 beads.   I find 6 more.   How many beads do I have in all?

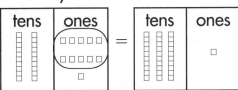

Regroup 10 ones as 1 ten.
I have 31 beads.

Solve each problem. Draw a picture to show the answer.

1.   I have   I find 6 more.   How many   I have _____
        36 beads.   beads in all?   beads.

2.   I have   I find 5 more.   How many   I have _____
        18 beads.   beads in all?   beads.

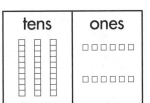

3.   I have   I find 6 more.   How many   I have _____
        28 beads.   beads in all?   beads.

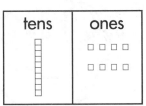
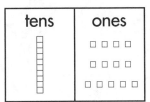
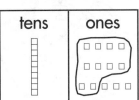

Adding a Two-Digit Number and a One-Digit Number

Use these steps to add and regroup.

1. Add the ones.
$6 + 8 = 14$ ones
Regroup before you
add the tens.

tens	ones
□	
2	6
+	8

2. Regroup 14 ones as
1 ten and 4 ones.
Write 4 ones in the sum.
Write 1 in the tens column.

tens	ones
1	
2	6
+	8
	4

3. Add the tens.
$1 + 2 = 3$ tens
Write 3 tens in
the sum.

The sum
is 34.

tens	ones
1	
2	6
+	8
3	4

Solve each problem.

1.
```
  35
+  6
____
```

2.
```
  52
+  8
____
```

3.
```
  19
+  5
____
```

4.
```
  66
+  7
____
```

5.
```
  24
+  9
____
```

6.
```
  85
+  6
____
```

7.
```
  74
+  8
____
```

8.
```
  48
+  3
____
```

9.
```
  17
+  5
____
```

10.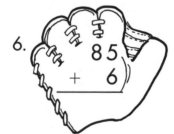
```
  31
+  9
____
```

11.
```
  87
+  7
____
```

12.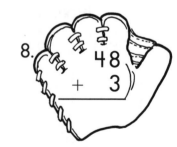
```
  56
+  8
____
```

Adding a Two-Digit Number and a One-Digit Number

Look at this!

tens	ones
☐1	
7	3
+	9
8	2

Solve the problems in order. Cross out each sum on the frozen ice-cream treats to see which treat is finished first.

1. 56
   + 8

2. 59
   + 9

3. 29
   + 4

4. 79
   + 2

5. 35
   + 6

6. 47
   + 5

7. 12
   + 8

8. 72
   + 9

9. 57
   + 7

68
81
52
81
64
55
61
30

10. 44
    + 6

11. 49
    + 6

12. 63
    + 8

13. 54
    + 7

11. 25
    + 5

12. 82
    + 8

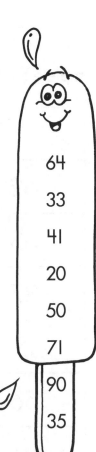

64
33
41
20
50
71
90
35

Adding Multiples of 10

Add a ten block to each group. Write the number sentence.

1. ▯▯▯▯▯ + ▯

___ + ___ = ___

2. ▯▯▯▯▯▯ + ▯

___ + ___ = ___

3. ▯▯▯▯▯▯▯ + ▯

___ + ___ = ___

4. ▯ + ▯

___ + ___ = ___

5. ▯▯ + ▯

___ + ___ = ___

6. ▯▯▯▯▯▯ + ▯

___ + ___ = ___

Complete the table by adding 10 to each number on the left.

+ 10	
62	
33	
87	
5	
56	
21	

Adding Multiples of 10

Which bowling pin will fall next? To find out, solve the problems in order. Cross off each sum on the bowling pins. The bowling pin with all of the numbers crossed off is the next one to fall. Color it.

1. 43
   + 10

2. 16
   + 10

3. 71
   + 10

4. 24
   + 10

5. 85
   + 10

6. 39
   + 10

7. 15
   + 10

8. 46
   + 10

9. 57
   + 10

10. 63
    + 10

11. 24
    + 10

12. 73
    + 10

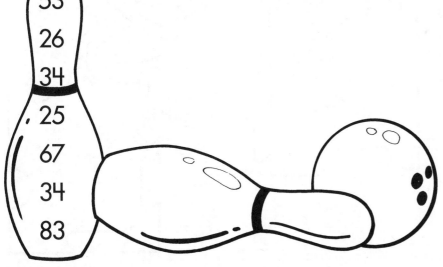

53
26
34
25
67
34
83

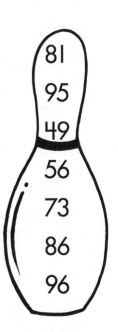

81
95
49
56
73
86
96

Adding Multiples of 10

Write each missing addend.

1.
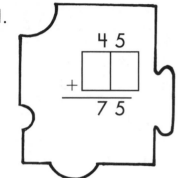

$$
\begin{array}{r}
4\ 5 \\
+\ \boxed{} \\
\hline
7\ 5
\end{array}
$$

2.
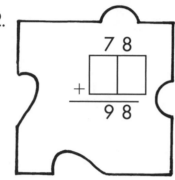

$$
\begin{array}{r}
7\ 8 \\
+\ \boxed{} \\
\hline
9\ 8
\end{array}
$$

3.
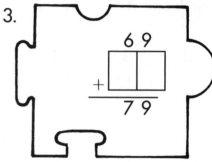

$$
\begin{array}{r}
6\ 9 \\
+\ \boxed{} \\
\hline
7\ 9
\end{array}
$$

4.
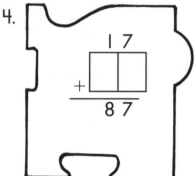

$$
\begin{array}{r}
1\ 7 \\
+\ \boxed{} \\
\hline
8\ 7
\end{array}
$$

5.
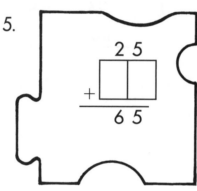

$$
\begin{array}{r}
2\ 5 \\
+\ \boxed{} \\
\hline
6\ 5
\end{array}
$$

6.
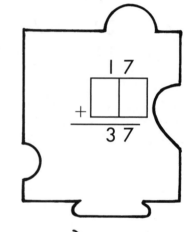

$$
\begin{array}{r}
1\ 7 \\
+\ \boxed{} \\
\hline
3\ 7
\end{array}
$$

7.
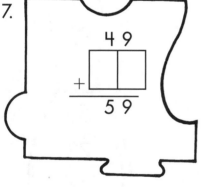

$$
\begin{array}{r}
4\ 9 \\
+\ \boxed{} \\
\hline
5\ 9
\end{array}
$$

8.
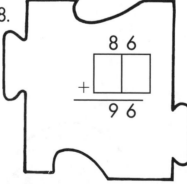

$$
\begin{array}{r}
8\ 6 \\
+\ \boxed{} \\
\hline
9\ 6
\end{array}
$$

Finding 10 More/10 Less

Write the number that is 10 more.

1.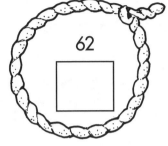
62

2.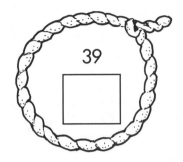
39

3.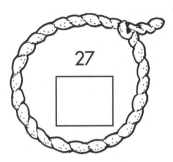
27

4.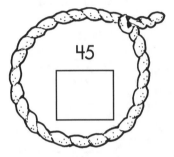
45

5.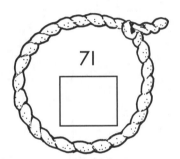
71

6.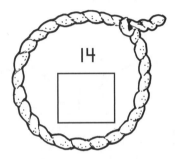
14

Write the number that is 10 less.

7.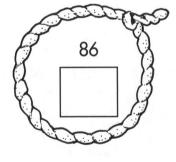
86

8.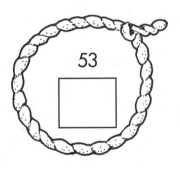
53

9.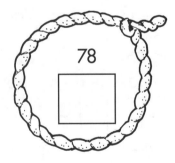
78

10.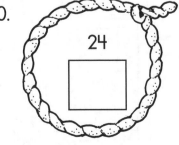
24

11.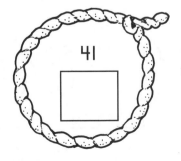
41

12.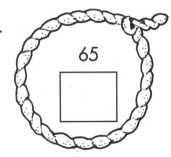
65

Finding 10 More/10 Less

Look at each number. Write the number that is 10 more.

53 49 21 70

34 72 38 16 25

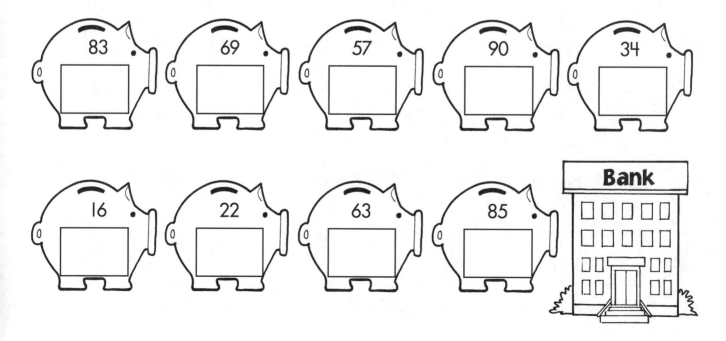

Look at each number. Write the number that is 10 less.

83 69 57 90 34

16 22 63 85 Bank

Finding 10 More/10 Less

Find two numbers around the square that have a difference of 10. Connect their dots with a straight line. Repeat until each dot has been connected to another.

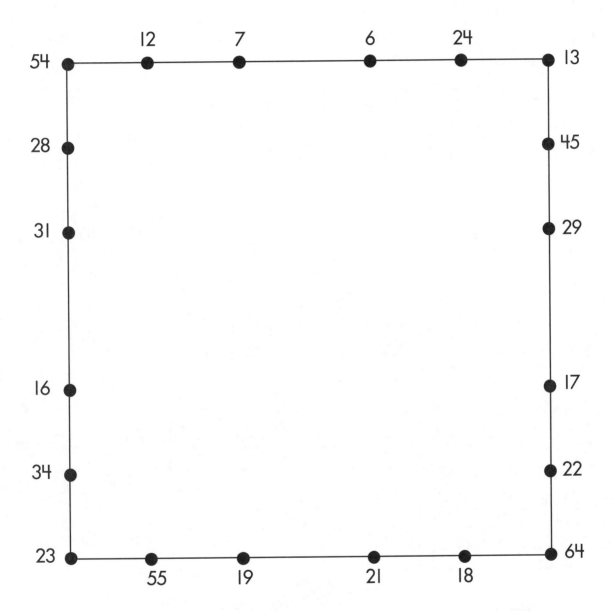

Subtracting Multiples of 10

Take 10 seeds away from each picture. Write the number.

1.

   35 − 10 = _____

2.

   59 − 10 = _____

3.

   18 − 10 = _____

4.

   62 − 10 = _____

5.

   70 − 10 = _____

6.

   47 − 10 = _____

Subtracting Multiples of 10

Solve each problem.

1. $\begin{array}{r} 63 \\ -\ 10 \\ \hline \end{array}$
2. $\begin{array}{r} 80 \\ -\ 10 \\ \hline \end{array}$
3. $\begin{array}{r} 75 \\ -\ 10 \\ \hline \end{array}$
4. $\begin{array}{r} 79 \\ -\ 10 \\ \hline \end{array}$
5. $\begin{array}{r} 38 \\ -\ 10 \\ \hline \end{array}$
6. $\begin{array}{r} 93 \\ -\ 10 \\ \hline \end{array}$

7. $\begin{array}{r} 67 \\ -\ 10 \\ \hline \end{array}$
8. $\begin{array}{r} 83 \\ -\ 10 \\ \hline \end{array}$
9. $\begin{array}{r} 77 \\ -\ 10 \\ \hline \end{array}$
10. $\begin{array}{r} 76 \\ -\ 10 \\ \hline \end{array}$
11. $\begin{array}{r} 59 \\ -\ 10 \\ \hline \end{array}$
12. $\begin{array}{r} 57 \\ -\ 10 \\ \hline \end{array}$

13. $\begin{array}{r} 56 \\ -\ 10 \\ \hline \end{array}$
14. $\begin{array}{r} 63 \\ -\ 10 \\ \hline \end{array}$
15. $\begin{array}{r} 48 \\ -\ 10 \\ \hline \end{array}$
16. $\begin{array}{r} 86 \\ -\ 10 \\ \hline \end{array}$
17. $\begin{array}{r} 40 \\ -\ 10 \\ \hline \end{array}$
18. $\begin{array}{r} 43 \\ -\ 10 \\ \hline \end{array}$

1.NBT.C.6

Subtracting Multiples of 10

Solve each problem.

1. $\begin{array}{r} 77 \\ - 40 \\ \hline \end{array}$

2. $\begin{array}{r} 55 \\ - 20 \\ \hline \end{array}$

3. $\begin{array}{r} 98 \\ - 50 \\ \hline \end{array}$

4. $\begin{array}{r} 29 \\ - 10 \\ \hline \end{array}$

5. $\begin{array}{r} 60 \\ - 50 \\ \hline \end{array}$

6. $\begin{array}{r} 69 \\ - 20 \\ \hline \end{array}$

7. $\begin{array}{r} 45 \\ - 20 \\ \hline \end{array}$

8. $\begin{array}{r} 78 \\ - 60 \\ \hline \end{array}$

9. $\begin{array}{r} 86 \\ - 80 \\ \hline \end{array}$

10. $\begin{array}{r} 39 \\ - 10 \\ \hline \end{array}$

11. $\begin{array}{r} 86 \\ - 20 \\ \hline \end{array}$

12. $\begin{array}{r} 59 \\ - 50 \\ \hline \end{array}$

13. $\begin{array}{r} 72 \\ - 30 \\ \hline \end{array}$

14. $\begin{array}{r} 93 \\ - 80 \\ \hline \end{array}$

15. $\begin{array}{r} 26 \\ - 10 \\ \hline \end{array}$

16. $\begin{array}{r} 98 \\ - 10 \\ \hline \end{array}$

17. $\begin{array}{r} 67 \\ - 30 \\ \hline \end{array}$

18. $\begin{array}{r} 39 \\ - 10 \\ \hline \end{array}$

19. $\begin{array}{r} 77 \\ - 10 \\ \hline \end{array}$

20. $\begin{array}{r} 87 \\ - 10 \\ \hline \end{array}$

21. $\begin{array}{r} 63 \\ - 30 \\ \hline \end{array}$

22. $\begin{array}{r} 82 \\ - 70 \\ \hline \end{array}$

23. $\begin{array}{r} 74 \\ - 50 \\ \hline \end{array}$

24. $\begin{array}{r} 83 \\ - 40 \\ \hline \end{array}$

Comparing Lengths

Color the longest object in each set red. Color the shortest object in each set blue.

1. Bolts

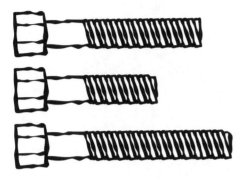

2. Screws

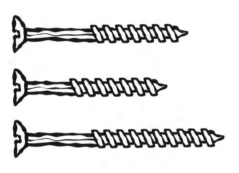

3. Nails

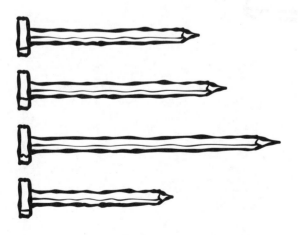

Comparing Lengths

Each clown's shoe is a different length. Write the clowns' names in order of the lengths of their shoes from shortest to longest.

You

Sou

Bou

Zou

_____   _____   _____   _____

shortest shoe   longest shoe

Ned

Ted

Hed

Zed

Red

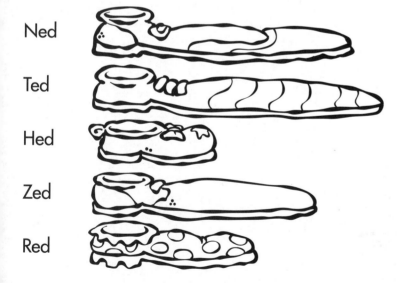

_____   _____   _____   _____   _____

shortest shoe   longest shoe

Comparing Lengths

Each clown's hat has a letter. Look at the heights of the hats and answer the questions using the letters for the hats.

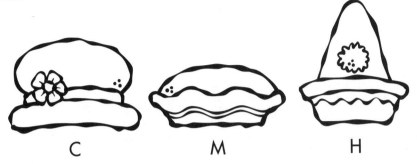

C   M   H

1. Which hat is the tallest? _____

2. Which hat is the shortest? _____

3. Write the hats' letter names in order from shortest to tallest.

_____   _____   _____

4. Circle the sentences that are true about the hats. Remember: > means greater (taller) than, < means less (shorter) than, and = means equal to.

C > M	H < C	M > H	C = H
H > M	M < H	C < M	H > C

Z   A   N   Y

5. Write the hats' letter names from tallest to shortest.

_____   _____   _____   _____

6. Circle the sentences that are true about the hats.

Z > Y	Z > A	A > Y	A < Y	A = N
N < A	Y > Z	N > Z	A < Z	Z < N

Measuring in Units

Write how long each object is in units.

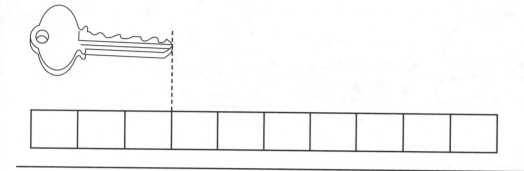

_____ units

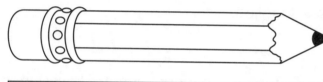

_____ units

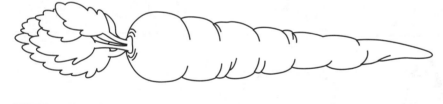

_____ units

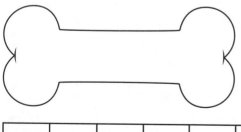

_____ units

Measuring in Units

Write how long each object is in units.

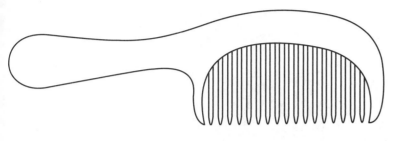

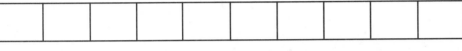

_____ units

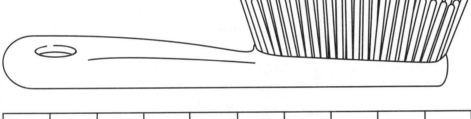

_____ units

_____ units

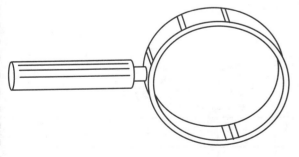

_____ units

Measuring in Units

Count how many cubes long each object is. Write the number.

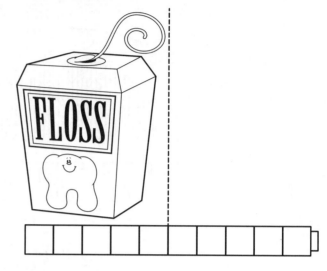

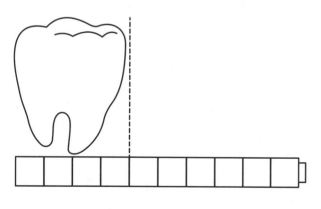

_____ cubes

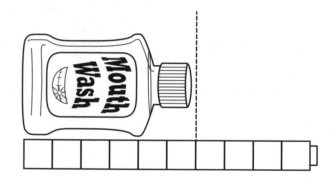

_____ cubes

_____ cubes

_____ cubes

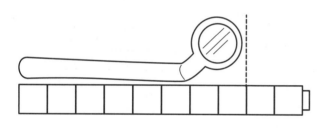

_____ cubes

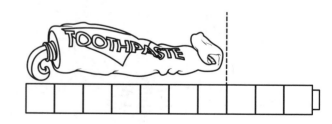

_____ cubes

Time to the Hour

The hour hand is the short hand on a clock.
It shows which hour it is.

6:00

Write the correct time under each clock.

___ ___ : ___ ___

___ ___ : ___ ___

___ ___ : ___ ___

___ ___ : ___ ___

___ ___ : ___ ___

___ ___ : ___ ___

Time to the Hour

Draw hands on each clock to show the correct time.

12:00

2:00

9:00

6:00

10:00

7:00

1.MD.B.3

Time to the Hour

Write the correct time under each watch.

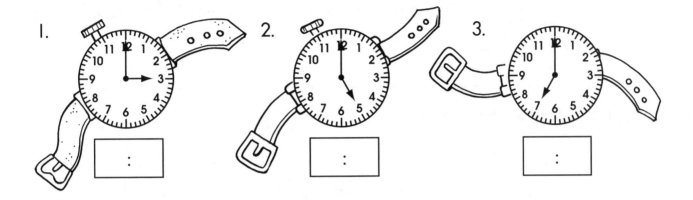

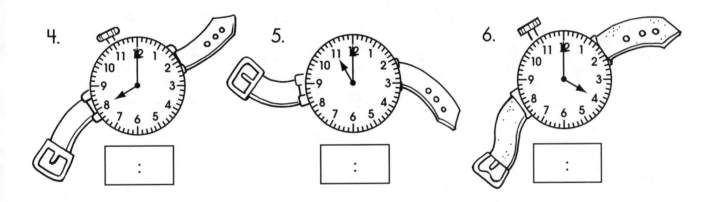

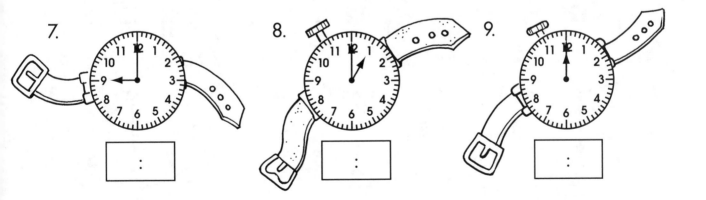

Time to the Half-Hour

 The minute hand is the longer hand. It is pointing to the 6. Thirty minutes have passed after 3:00. The time is 3:30.

Write the correct time under each clock.

____ : ____ ____

____ : ____ ____

____ ____ : ____ ____

____ : ____ ____

____ : ____ ____

____ : ____ ____

Time to the Half-Hour

Draw hands on each clock to show the correct time.

4:30

6:30

1:30

10:30

9:30

12:30

1.MD.B.3

Time to the Half-Hour

Draw the hands on each clock to show the correct time.

1.

7:30

2.

2:30

3.

8:30

4.

4:30

5.

6:30

6.

11:30

7

5:30

8.

10:30

9.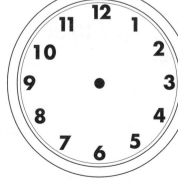

12:30

Analyzing Data

Ms. Smith's first-grade class made a tally chart about their pets. Each student made only one tally. Use the chart to answer the questions.

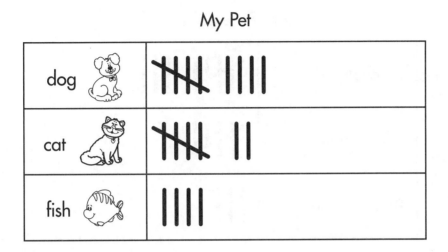

My Pet

1. How many students have a dog? _____

2. How many more students have a dog than a cat? _____

3. How many students have a fish? _____

4. How many students are in Ms. Smith's class? _____

Analyzing Data

Mrs. Mason's class has earned an extra recess! They voted to decide what to do. Use the tally chart to answer the question.

Extra Recess Games

Kickball								
Four square					
Board games				

1. Which game was chosen the least? _____

2. How many students are in Mrs. Mason's class? _____

3. Which game was chosen the most? _____

4. How many more votes did four square get than board games? _____

Analyzing Data

Ask 10 people which drink they like best. In the table below, make a tally mark beside the drink each one likes. Use the chart to answer the questions.

Favorite Kind of Drink

1. How many people like milk? _____

2. How many people like juice? _____

3. Do more people like water or milk? _____

4. How many people like water and juice in all? _____

Attributes of Two-Dimensional Shapes

Cut out the shapes below. Glue each shape to its matching shape. Write the number of sides and corners.

Sides

Corners

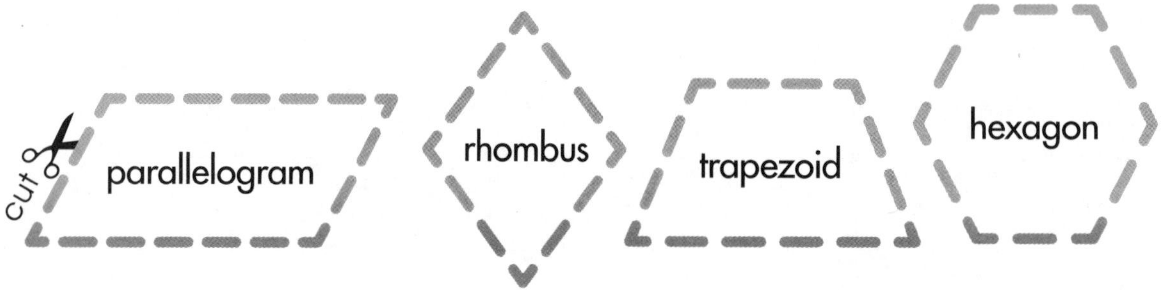

✂ cut

parallelogram

rhombus

trapezoid

hexagon

Attributes of Two-Dimensional Shapes

Look at the picture of a log cabin. How many of each shape can you find in the picture?

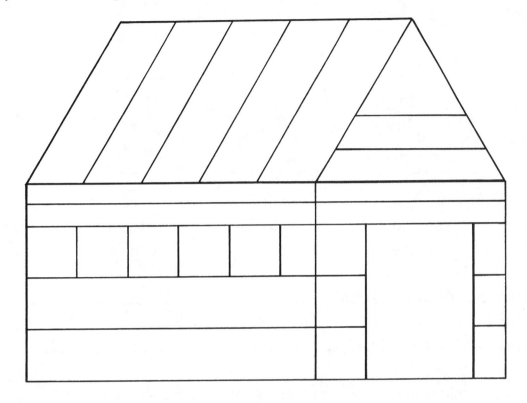

How many △ ? _____   How many ▱ ? _____

How many ☐ ? _____   How many ▭ ? _____   How many ⬟ ? _____

How many sides does each shape have?

△ _____   ▱ _____

☐ _____   ▭ _____   ⬭ _____

How many corners does each shape have?

△ _____   ▱ _____

☐ _____   ▭ _____   ⬭ _____

Attributes of Two-Dimensional Shapes

Jan made cookies with her new cookie cutters. Count how many sides each cookie has. Then, answer the questions below.

_____ sides   _____ sides   _____ sides   _____ sides   _____ sides

H   S   T   O   P

hexagon   square   triangle   octagon   pentagon
cookie   cookie   cookie   cookie   cookie

1. Which cookie has the fewest sides? _____

2. Which cookie has the most sides? _____

3. Write the cookie letters in order of the number of sides the shapes have.

_____   _____   _____   _____   _____

fewest sides   most sides

4. How many more sides does O have than H? _____

5. How many more sides does H have than S? _____

6. How many fewer sides does T have than P? _____

7. How many fewer sides does S have than O? _____

8. How many more sides would P need to become O? _____

Attributes of Three-Dimensional Shapes

The flat side of a solid
shape is called a face.

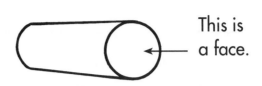

This is
a face.

Complete the table.

Spatial Shape	Number of Faces
▲	
⬜	
▮	
▱	
⬤	

Attributes of Three-Dimensional Shapes

Complete the table.

	Number of Faces	Number of Edges	Number of Vertices (Corners)

Attributes of Three-Dimensional Shapes

Cut out the shapes. Glue each shape in the correct box.

Has More Than Two Faces	Has Two or Less Faces

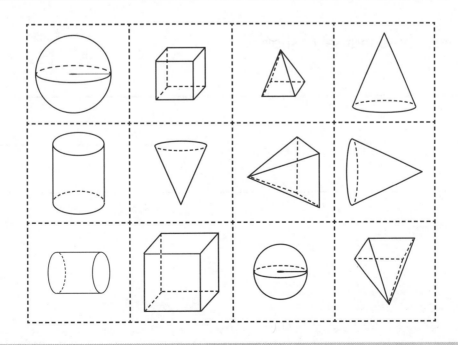

Composing Shapes

Look at the figure. Circle the set of shapes that form this figure when put together.

A   B   C

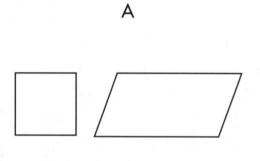 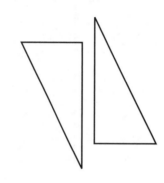 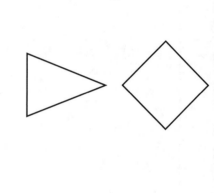

Tell or write why you chose that set of shapes.

_____

_____

_____

Composing Shapes

Look at the figure. Circle the set of shapes that form this figure when put together.

A	B	C

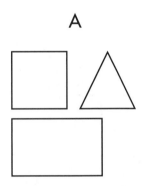

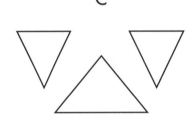

Tell or write why you chose that set of shapes.

_____

_____

_____

Composing Shapes

Look at the figure. Circle the set of shapes that form this figure when put together.

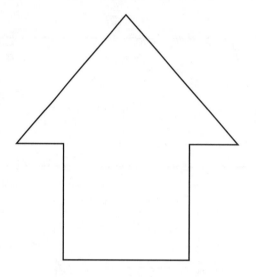

A

B

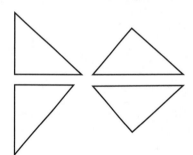

C

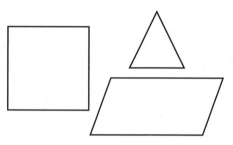

Tell or write why you chose that set of shapes.

_____

_____

_____

Partitioning Shapes

Color the shapes with two equal parts (halves) blue. Color the shapes with four equal parts (fourths) yellow.

1.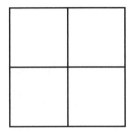

2.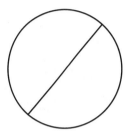

3.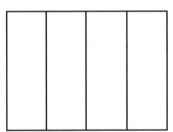

4.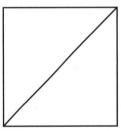

5.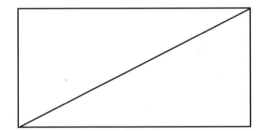

6.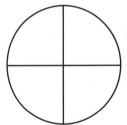

7.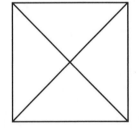

7.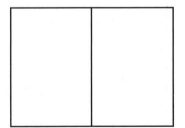

Partitioning Shapes

Look at each shape. Circle the word that describes the number of equal parts.

1. 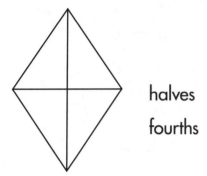 halves
   fourths

2. 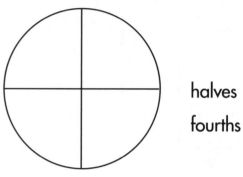 halves
   fourths

3. 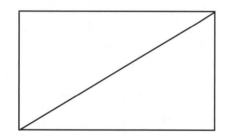 halves
   fourths

4. 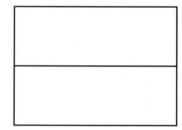 halves
   fourths

5. 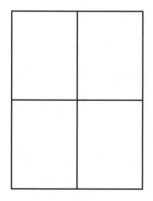 halves
   fourths

6. 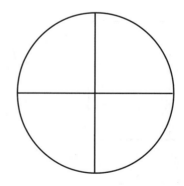 halves
   fourths

Partitioning Shapes

Draw lines to divide each shape into the correct number of equal parts.

1.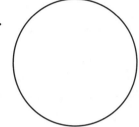

halves

2.

fourths

3.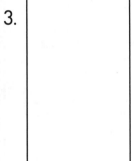

halves

4.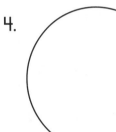

fourths

Color each picture to show the fraction.

5.

two-fourths

6.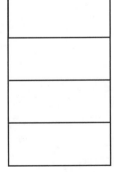

one-fourth

Answer Key

Name _____ 1.OA.A.1

Solving Word Problems: Adding To

For each story, draw a picture and write a number sentence.

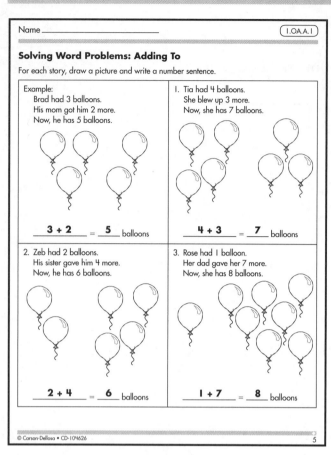

Example:
Brad had 3 balloons.
His mom got him 2 more.
Now, he has 5 balloons.

__3 + 2__ = __5__ balloons

1. Tia had 4 balloons.
She blew up 3 more.
Now, she has 7 balloons.

__4 + 3__ = __7__ balloons

2. Zeb had 2 balloons.
His sister gave him 4 more.
Now, he has 6 balloons.

__2 + 4__ = __6__ balloons

3. Rose had 1 balloon.
Her dad gave her 7 more.
Now, she has 8 balloons.

__1 + 7__ = __8__ balloons

© Carson-Dellosa • CD-104626   5

Name _____ 1.OA.A.1

Solving Word Problems: Adding To

For each story, draw a picture and write a number sentence.

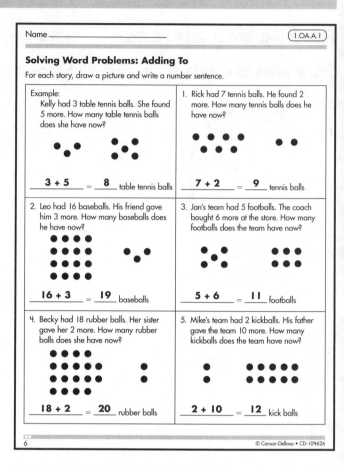

Example:
Kelly had 3 table tennis balls. She found 5 more. How many table tennis balls does she have now?

__3 + 5__ = __8__ table tennis balls

1. Rick had 7 tennis balls. He found 2 more. How many tennis balls does he have now?

__7 + 2__ = __9__ tennis balls

2. Leo had 16 baseballs. His friend gave him 3 more. How many baseballs does he have now?

__16 + 3__ = __19__ baseballs

3. Jan's team had 5 footballs. The coach bought 6 more at the store. How many footballs does the team have now?

__5 + 6__ = __11__ footballs

4. Becky had 18 rubber balls. Her sister gave her 2 more. How many rubber balls does she have now?

__18 + 2__ = __20__ rubber balls

5. Mike's team had 2 kickballs. His father gave the team 10 more. How many kickballs does the team have now?

__2 + 10__ = __12__ kick balls

6   © Carson-Dellosa • CD-104626

Name _____ 1.OA.A.1

Solving Word Problems: Adding To

For each number sentence, write an "adding to" story.

Example:
3 + 2 = 5   I had 3 purple markers. My friend gave me 2 more. Now, I have 5 purple markers.

1. 4 + 2 = 6   **Answers will vary.**

2. 7 + 4 = 11   **Answers will vary.**

3. 5 + 4 = 9   **Answers will vary.**

4. 16 + 3 = 19   **Answers will vary.**

© Carson-Dellosa • CD-104626   7

Name _____ 1.OA.A.1

Solving Word Problems: Taking From

For each story, draw a picture and write a number sentence.

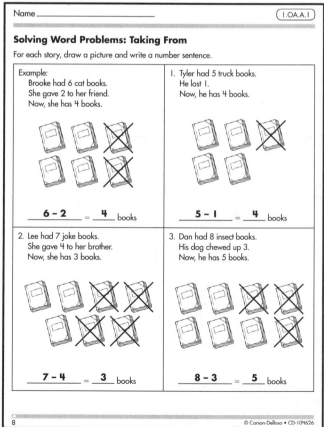

Example:
Brooke had 6 cat books.
She gave 2 to her friend.
Now, she has 4 books.

__6 – 2__ = __4__ books

1. Tyler had 5 truck books.
He lost 1.
Now, he has 4 books.

__5 – 1__ = __4__ books

2. Lee had 7 joke books.
She gave 4 to her brother.
Now, she has 3 books.

__7 – 4__ = __3__ books

3. Dan had 8 insect books.
His dog chewed up 3.
Now, he has 5 books.

__8 – 3__ = __5__ books

8   © Carson-Dellosa • CD-104626

© Carson-Dellosa • CD-104626

Answer Key

Name _____

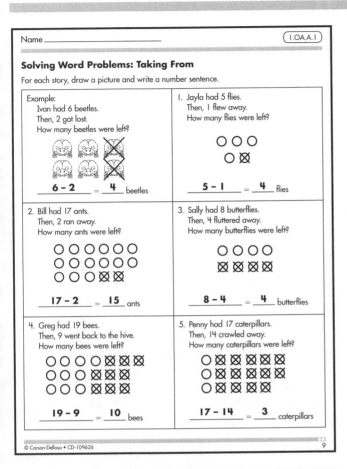

Solving Word Problems: Taking From

For each story, draw a picture and write a number sentence.

Example:
Ivan had 6 beetles.
Then, 2 got lost.
How many beetles were left?

__6 - 2__ = __4__ beetles

1. Jayla had 5 flies.
Then, 1 flew away.
How many flies were left?

__5 - 1__ = __4__ flies

2. Bill had 17 ants.
Then, 2 ran away.
How many ants were left?

__17 - 2__ = __15__ ants

3. Sally had 8 butterflies.
Then, 4 fluttered away.
How many butterflies were left?

__8 - 4__ = __4__ butterflies

4. Greg had 19 bees.
Then, 9 went back to the hive.
How many bees were left?

__19 - 9__ = __10__ bees

5. Penny had 17 caterpillars.
Then, 14 crawled away.
How many caterpillars were left?

__17 - 14__ = __3__ caterpillars

© Carson-Dellosa • CD-104626   9

Name _____   1.OA.A.1

Solving Word Problems: Taking From

For each number sentence, write a "taking from" story.

Example:
9 − 2 = 7   I had 9 computer games. I gave my friend 2 games. Now, I have 7 computer games.

1. 16 − 6 = 10   **Answers will vary.**

2. 12 − 7 = 5   **Answers will vary.**

3. 9 − 3 = 6   **Answers will vary.**

4. 18 − 10 = 8   **Answers will vary.**

10   © Carson-Dellosa • CD-104626

Name _____   1.OA.A.1

Solving Word Problems: Putting Together/Taking Apart

For each story, draw a picture. Write the answer on the line.

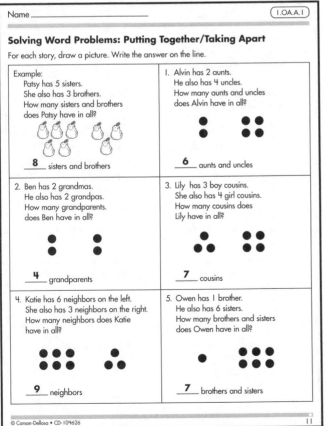

Example:
Patsy has 5 sisters.
She also has 3 brothers.
How many sisters and brothers does Patsy have in all?

__8__ sisters and brothers

1. Alvin has 2 aunts.
He also has 4 uncles.
How many aunts and uncles does Alvin have in all?

__6__ aunts and uncles

2. Ben has 2 grandmas.
He also has 2 grandpas.
How many grandparents does Ben have in all?

__4__ grandparents

3. Lily has 3 boy cousins.
She also has 4 girl cousins.
How many cousins does Lily have in all?

__7__ cousins

4. Katie has 6 neighbors on the left.
She also has 3 neighbors on the right.
How many neighbors does Katie have in all?

__9__ neighbors

5. Owen has 1 brother.
He also has 6 sisters.
How many brothers and sisters does Owen have in all?

__7__ brothers and sisters

© Carson-Dellosa • CD-104626   11

Name _____   1.OA.A.1

Solving Word Problems: Putting Together/Taking Apart

Write a number sentence to solve each problem.

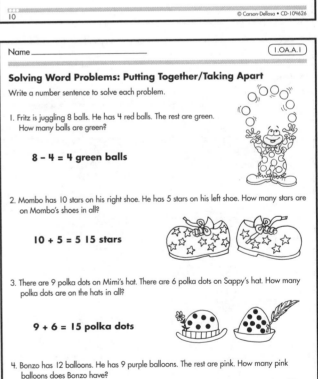

1. Fritz is juggling 8 balls. He has 4 red balls. The rest are green. How many balls are green?

8 − 4 = 4 green balls

2. Mombo has 10 stars on his right shoe. He has 5 stars on his left shoe. How many stars are on Mombo's shoes in all?

10 + 5 = 5 15 stars

3. There are 9 polka dots on Mimi's hat. There are 6 polka dots on Sappy's hat. How many polka dots are on the hats in all?

9 + 6 = 15 polka dots

4. Bonzo has 12 balloons. He has 9 purple balloons. The rest are pink. How many pink balloons does Bonzo have?

12 − 9 = 3 pink balloons

12   © Carson-Dellosa • CD-104626

Answer Key

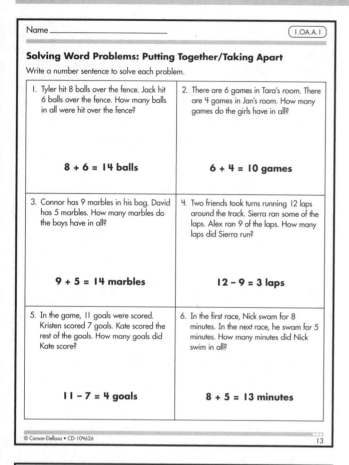

Name _____ (1.OA.A.1)

Solving Word Problems: Putting Together/Taking Apart

Write a number sentence to solve each problem.

1. Tyler hit 8 balls over the fence. Jack hit 6 balls over the fence. How many balls in all were hit over the fence?

8 + 6 = 14 balls

2. There are 6 games in Tara's room. There are 4 games in Jan's room. How many games do the girls have in all?

6 + 4 = 10 games

3. Connor has 9 marbles in his bag. David has 5 marbles. How many marbles do the boys have in all?

9 + 5 = 14 marbles

4. Two friends took turns running 12 laps around the track. Sierra ran some of the laps. Alex ran 9 of the laps. How many laps did Sierra run?

12 – 9 = 3 laps

5. In the game, 11 goals were scored. Kristen scored 7 goals. Kate scored the rest of the goals. How many goals did Kate score?

11 – 7 = 4 goals

6. In the first race, Nick swam for 8 minutes. In the next race, he swam for 5 minutes. How many minutes did Nick swim in all?

8 + 5 = 13 minutes

© Carson-Dellosa • CD-104626   13

Name _____ (1.OA.A.1)

Solving Word Problems: Comparing

Bill and Becky counted some things they own and made graphs to compare their totals. Looks at the graphs and answer the questions. Then, write a number sentence to show the comparison.

1. Board Games
Who has more? **Bill**
How many more? **2**
7 – 5 = 2 games

2. Hats
Who has more? **Becky**
How many more? **1**
5 – 4 = 1 hat

3. Balls
Who has more? **Bill**
How many more? **2**
5 – 3 = 2 balls

4. Video Games
Who has more? **Becky**
How many more? **3**
7 – 4 = 3 games

5. Hamsters
Who has more? **Becky**
How many more? **3**
6 – 3 = 3 hamsters

6. Posters
Who has more? **Bill**
How many more? **4**
7 – 3 = 4 posters

14   © Carson-Dellosa • CD-104626

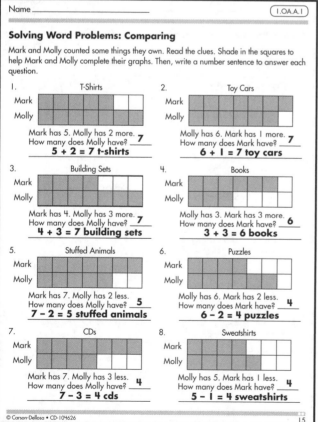

Name _____ (1.OA.A.1)

Solving Word Problems: Comparing

Mark and Molly counted some things they own. Read the clues. Shade in the squares to help Mark and Molly complete their graphs. Then, write a number sentence to answer each question.

1. T-Shirts
Mark has 5. Molly has 2 more. How many does Molly have? **7**
5 + 2 = 7 t-shirts

2. Toy Cars
Molly has 6. Mark has 1 more. How many does Mark have? **7**
6 + 1 = 7 toy cars

3. Building Sets
Mark has 4. Molly has 3 more. How many does Molly have? **7**
4 + 3 = 7 building sets

4. Books
Molly has 3. Mark has 3 more. How many does Mark have? **6**
3 + 3 = 6 books

5. Stuffed Animals
Mark has 7. Molly has 2 less. How many does Molly have? **5**
7 – 2 = 5 stuffed animals

6. Puzzles
Molly has 6. Mark has 2 less. How many does Mark have? **4**
6 – 2 = 4 puzzles

7. CDs
Mark has 7. Molly has 3 less. How many does Molly have? **4**
7 – 3 = 4 cds

8. Sweatshirts
Molly has 5. Mark has 1 less. How many does Mark have? **4**
5 – 1 = 4 sweatshirts

© Carson-Dellosa • CD-104626   15

Name _____ (1.OA.A.1)

Solving Word Problems: Comparing

Luke and Lily counted the stickers in their collections. Draw a picture and write a number sentence to solve each problem.

1. **Drawings will vary.**
3 + 2 = 5 fish stickers
Luke has 3 fish stickers. Lily has 2 more. How many does Lily have? **5**

2. **Drawings will vary.**
6 – 2 = 4 dog stickers
Luke has 6 dog stickers. Lily has 2 less. How many does Lily have? **4**

3. **Drawings will vary.**
4 + 3 = 7 dinosaur stickers
Luke has 4 dinosaur stickers. Lily has 3 more. How many does Lily have? **7**

4. **Drawings will vary.**
6 – 3 = 3 space stickers
Luke has 6 space stickers. Lily has 3 less. How many does Lily have? **3**

5. **Drawings will vary.**
6 + 1 = 7 cat stickers
Lily has 6 cat stickers. Luke has 1 more. How many does Luke have? **7**

6. **Drawings will vary.**
7 – 2 = 5 elephant stickers
Lily has 7 elephant stickers. Luke has 2 less. How many does Luke have? **5**

7. **Drawings will vary.**
4 – 3 = 1 ocean sticker
Lily has 4 ocean stickers. Luke has 3 less. How many does Luke have? **1**

8. **Drawings will vary.**
5 + 1 = 6 bird stickers
Lily has 5 bird stickers. Luke has 1 more. How many does Luke have? **6**

16   © Carson-Dellosa • CD-104626

Answer Key

Solving Word Problems: Three Numbers with Sums to 20

When finding the sum of three addends, look for a fact you know.

6
2 > 8
$+ 4$
$8 + 4 = 12$

6
2 > 6
$+ 4$
$6 + 6 = 12$

6
2 > 10
$+ 4$
$10 + 2 = 12$

Solve each problem. Circle the numbers you would add first.

1. ③
⑤
+ 4
12

2. ③
⑧
+ 6
17

3. ⑦
①
+ 5
13

4. ⑥
④
+ 1
11

Answers will vary for circling numbers.

5. ⑨
②
+ 3
14

6. ④
5
+④
13

7. ②
①
+ 6
9

8. ③
③
+ 5
11

9. David ate 1 banana, 8 grapes, and 9 blueberries for breakfast. How many pieces of fruit did he eat in all?
1 + 8 + 9 = 18 pieces of fruit

10. Myra drew 5 rabbits, 5 chicks, and 3 sheep in her picture. How many animals did she draw in all?
5 + 5 + 3 = 13 animals

Solving Word Problems: Three Numbers with Sums to 20

When finding the sum of three addends, look for a fact you know.

6
2 > 8
$+ 4$
$8 + 4 = 12$

6
2 > 6
$+ 4$
$6 + 6 = 12$

6
2 > 10
$+ 4$
$10 + 2 = 12$

Solve each problem. Circle the numbers you would add first.

1. 2
①
+⑨
12

2. ⑧
②
+ 7
17

3. 1
⑦
+③
11

4. ⑤
⑥
+ 2
13

Answers will vary for circling numbers.

5. Oliver found 2 pennies, 5 dimes, and 8 nickels in his pocket. How many coins did he find in all?
2 + 5 + 8 = 15 coins

6. Casey counted 7 red birds, 3 blue birds, and 6 yellow birds on her bird feeder. How many birds did she count in all?
7 + 3 + 6 = 16 birds

7. Erin had 6 marbles, Brad had 4 marbles, and Chelsea had 4 marbles. How many marbles did the children have in all?
6 + 4 + 4 = 14 marbles

8. Pablo swam 7 laps in the morning. He swam 4 more laps after lunch. He swam 2 more laps after dinner. How many laps did Pablo swim in all?
7 + 4 + 2 = 13 laps

Solving Word Problems: Three Numbers with Sums to 20

Solve each problem. Circle the numbers you would add first.

1. ⑥+④+ 2 = **12**

2. ①+①+ 1 = **3**

3. 5 +④+⑥= **15**

4. ⑦+③+ 2 = **12**

5. 4 +③+⑦= **14**

6. ④+④+ 8 = **16**

7. 6 +②+⑧= **16**

8. ②+②+ 4 = **8**

9. 9 +④+③= **16**

10. ③+ 4 +③= **10**

11. Alicia baked 7 chocolate chip cookies, 5 peanut butter cookies, and 6 oatmeal cookies. How many cookies did Alicia bake in all?
7 + 5 + 6 = 18 cookies

12. Gavin bought 3 packs of gum. Each pack has 5 sticks of gum. How many sticks of gum does Gavin have in all?
5 + 5 + 5 = 15 sticks of gum

13. In the pond, 3 turtles, 6 ducks, and 2 swans are swimming. How many animals are swimming in the pond in all?
3 + 6 + 2 = 11 animals

14. Jill has 7 pink flowers, 6 orange flowers, and 6 red flowers in a vase. How many flowers does Jill have in the vase in all?
7 + 6 + 6 = 19 flowers

Properties of Operations: Commutative Property

Count the dots on each domino. Solve each problem.

1 + 2 = **3**

2 + 1 = **3**

4 + 6 = **10**

6 + 4 = **10**

6 + 3 = **9**

3 + 6 = **9**

What do you notice about the two problems in each row?

Answers will vary but should include that no matter the order of the numbers, the sum, or answer, is the same.

Answer Key

Name

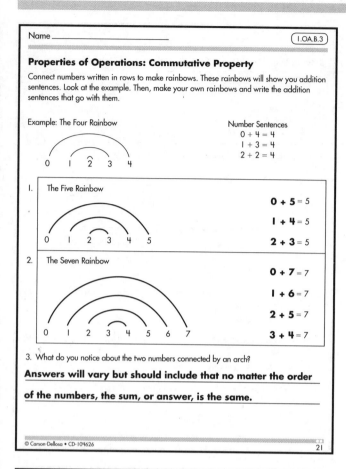

I.OA.B.3

Properties of Operations: Commutative Property

Connect numbers written in rows to make rainbows. These rainbows will show you addition sentences. Look at the example. Then, make your own rainbows and write the addition sentences that go with them.

Example: The Four Rainbow

Number Sentences
$0 + 4 = 4$
$1 + 3 = 4$
$2 + 2 = 4$

0 1 2 3 4

1. The Five Rainbow

0 + 5 = 5
1 + 4 = 5
2 + 3 = 5

0 1 2 3 4 5

2. The Seven Rainbow

0 + 7 = 7
1 + 6 = 7
2 + 5 = 7
3 + 4 = 7

0 1 2 3 4 5 6 7

3. What do you notice about the two numbers connected by an arch?

Answers will vary but should include that no matter the order of the numbers, the sum, or answer, is the same.

© Carson-Dellosa • CD-104626   21

Name   I.OA.B.3

Properties of Operations: Commutative Property

Make rainbows for these numbers and write the addition sentences that go with them. The first one has been started for you.

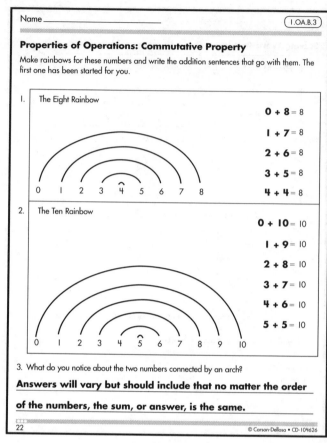

1. The Eight Rainbow

0 + 8 = 8
1 + 7 = 8
2 + 6 = 8
3 + 5 = 8
4 + 4 = 8

0 1 2 3 4 5 6 7 8

2. The Ten Rainbow

0 + 10 = 10
1 + 9 = 10
2 + 8 = 10
3 + 7 = 10
4 + 6 = 10
5 + 5 = 10

0 1 2 3 4 5 6 7 8 9 10

3. What do you notice about the two numbers connected by an arch?

Answers will vary but should include that no matter the order of the numbers, the sum, or answer, is the same.

22   © Carson-Dellosa • CD-104626

Name   I.OA.B.3

Properties of Operations: Associative Property

Solve each problem. Circle the two problems in each row that have the same sum.

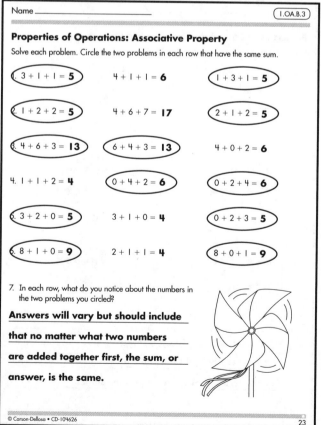

1. (3 + 1 + 1 = **5**)   4 + 1 + 1 = **6**   (1 + 3 + 1 = **5**)

2. (1 + 2 + 2 = **5**)   4 + 6 + 7 = **17**   (2 + 1 + 2 = **5**)

3. (4 + 6 + 3 = **13**)   (6 + 4 + 3 = **13**)   4 + 0 + 2 = **6**

4. 1 + 1 + 2 = **4**   (0 + 4 + 2 = **6**)   (0 + 2 + 4 = **6**)

5. (3 + 2 + 0 = **5**)   3 + 1 + 0 = **4**   (0 + 2 + 3 = **5**)

6. (8 + 1 + 0 = **9**)   2 + 1 + 1 = **4**   (8 + 0 + 1 = **9**)

7. In each row, what do you notice about the numbers in the two problems you circled?

Answers will vary but should include that no matter what two numbers are added together first, the sum, or answer, is the same.

© Carson-Dellosa • CD-104626   23

Name   I.OA.B.3

Properties of Operations: Associative Property

Solve each problem. Circle the two problems in each row that have the same sum.

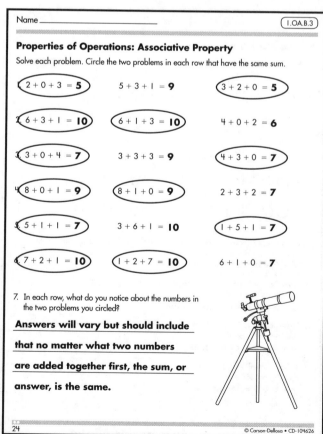

1. (2 + 0 + 3 = **5**)   5 + 3 + 1 = **9**   (3 + 2 + 0 = **5**)

2. (6 + 3 + 1 = **10**)   (6 + 1 + 3 = **10**)   4 + 0 + 2 = **6**

3. (3 + 0 + 4 = **7**)   3 + 3 + 3 = **9**   (4 + 3 + 0 = **7**)

4. (8 + 0 + 1 = **9**)   (8 + 1 + 0 = **9**)   2 + 3 + 2 = **7**

5. (5 + 1 + 1 = **7**)   3 + 6 + 1 = **10**   (1 + 5 + 1 = **7**)

6. (7 + 2 + 1 = **10**)   (1 + 2 + 7 = **10**)   6 + 1 + 0 = **7**

7. In each row, what do you notice about the numbers in the two problems you circled?

Answers will vary but should include that no matter what two numbers are added together first, the sum, or answer, is the same.

24   © Carson-Dellosa • CD-104626

Answer Key

Properties of Operations: Associative Property

Solve the problems in each row. Draw lines to connect the problems that have the same sum. The first one has been done for you.

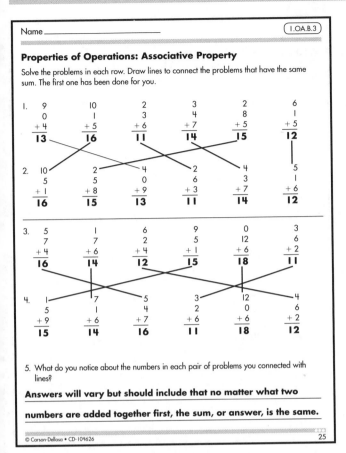

5. What do you notice about the numbers in each pair of problems you connected with lines?

Answers will vary but should include that no matter what two

numbers are added together first, the sum, or answer, is the same.

25

Understanding Subtraction as an Unknown Addend Problem

Draw a picture to solve each problem. Then, write an addition sentence and a subtraction sentence for each problem.

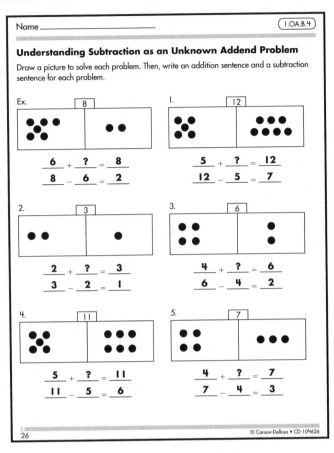

26

Understanding Subtraction as an Unknown Addend Problem

The Party Stuff Store counted up all of its party stuff and made some graphs to compare the totals. Find the difference for each graph. Write an addition sentence and a subtraction sentence about each difference. Circle the difference in each sentence.

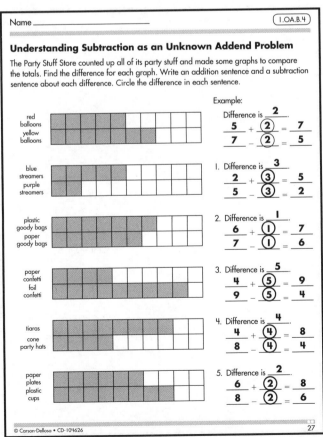

27

Understanding Subtraction as an Unknown Addend Problem

Write a subtraction sentence to solve for the missing number in each number sentence.

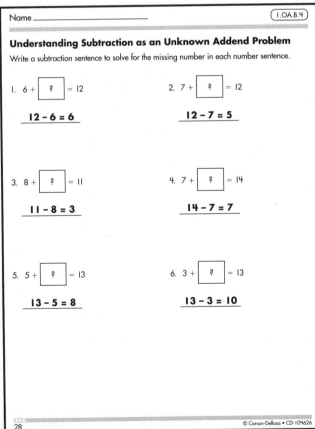

1. $6 + \boxed{?} = 12$

   12 − 6 = 6

2. $7 + \boxed{?} = 12$

   12 − 7 = 5

3. $8 + \boxed{?} = 11$

   11 − 8 = 3

4. $7 + \boxed{?} = 14$

   14 − 7 = 7

5. $5 + \boxed{?} = 13$

   13 − 5 = 8

6. $3 + \boxed{?} = 13$

   13 − 3 = 10

28

Answer Key

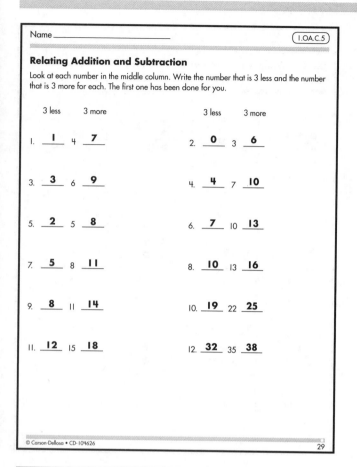

Name _____ 1.OA.C.5

Relating Addition and Subtraction

Look at each number in the middle column. Write the number that is 3 less and the number that is 3 more for each. The first one has been done for you.

3 less   3 more   3 less   3 more

1. **1** 4 **7**   2. **0** 3 **6**

3. **3** 6 **9**   4. **4** 7 **10**

5. **2** 5 **8**   6. **7** 10 **13**

7. **5** 8 **11**   8. **10** 13 **16**

9. **8** 11 **14**   10. **19** 22 **25**

11. **12** 15 **18**   12. **32** 35 **38**

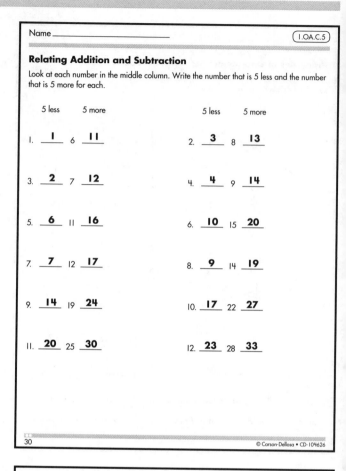

Name _____ 1.OA.C.5

Relating Addition and Subtraction

Look at each number in the middle column. Write the number that is 5 less and the number that is 5 more for each.

5 less   5 more   5 less   5 more

1. **1** 6 **11**   2. **3** 8 **13**

3. **2** 7 **12**   4. **4** 9 **14**

5. **6** 11 **16**   6. **10** 15 **20**

7. **7** 12 **17**   8. **9** 14 **19**

9. **14** 19 **24**   10. **17** 22 **27**

11. **20** 25 **30**   12. **23** 28 **33**

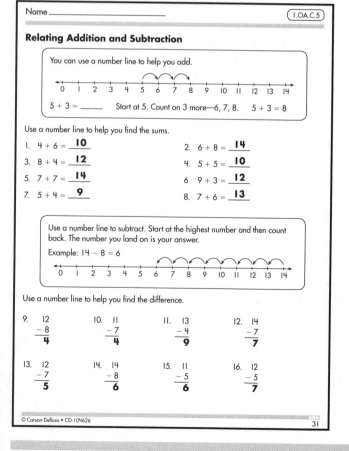

Name _____ 1.OA.C.5

Relating Addition and Subtraction

You can use a number line to help you add.

0 1 2 3 4 5 6 7 8 9 10 11 12 13 14

$5 + 3 =$ _____   Start at 5. Count on 3 more—6, 7, 8. $5 + 3 = 8$

Use a number line to help you find the sums.

1. $4 + 6 =$ **10**   2. $6 + 8 =$ **14**
3. $8 + 4 =$ **12**   4. $5 + 5 =$ **10**
5. $7 + 7 =$ **14**   6. $9 + 3 =$ **12**
7. $5 + 4 =$ **9**   8. $7 + 6 =$ **13**

Use a number line to subtract. Start at the highest number and then count back. The number you land on is your answer.

Example: $14 - 8 = 6$

0 1 2 3 4 5 6 7 8 9 10 11 12 13 14

Use a number line to help you find the difference.

9. 12 / −8 = **4**   10. 11 / −7 = **4**   11. 13 / −4 = **9**   12. 14 / −7 = **7**

13. 12 / −7 = **5**   14. 14 / −8 = **6**   15. 11 / −5 = **6**   16. 12 / −5 = **7**

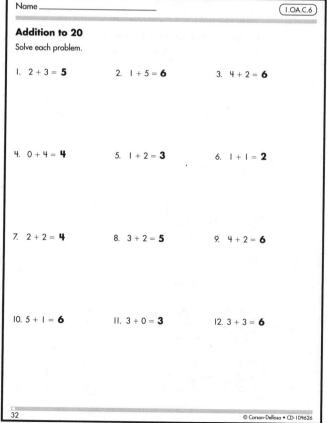

Name _____ 1.OA.C.6

Addition to 20

Solve each problem.

1. $2 + 3 =$ **5**   2. $1 + 5 =$ **6**   3. $4 + 2 =$ **6**

4. $0 + 4 =$ **4**   5. $1 + 2 =$ **3**   6. $1 + 1 =$ **2**

7. $2 + 2 =$ **4**   8. $3 + 2 =$ **5**   9. $4 + 2 =$ **6**

10. $5 + 1 =$ **6**   11. $3 + 0 =$ **3**   12. $3 + 3 =$ **6**

Answer Key

Name _____   1.OA.C.6

Addition to 20
Solve each problem.

1.	2.	3.	4.	5.
10 + 1 **11**	4 + 7 **11**	3 + 6 **9**	8 + 3 **11**	7 + 5 **12**

6.	7.	8.	9.	10.
5 + 7 **12**	10 + 2 **12**	3 + 7 **10**	4 + 8 **12**	9 + 2 **11**

11.	12.	13.	14.	15.
1 + 9 **10**	5 + 3 **8**	6 + 6 **12**	7 + 3 **10**	7 + 4 **11**

33

Name _____   1.OA.C.6

Addition to 20
Solve each problem.

1.	2.	3.	4.	5.
1 + 8 **9**	15 + 3 **18**	10 + 8 **18**	12 + 8 **20**	8 + 9 **17**

6.	7.	8.	9.	10.
9 + 3 **12**	11 + 2 **13**	11 + 7 **18**	12 + 5 **17**	8 + 2 **10**

11.	12.	13.	14.	15.
8 + 5 **13**	9 + 9 **18**	17 + 3 **20**	9 + 7 **16**	13 + 5 **18**

34

Name _____   1.OA.C.6

Subtraction from 20
Solve each problem.

1.	2.	3.	4.	5.
2 − 0 **2**	5 − 5 **0**	3 − 1 **2**	3 − 3 **0**	3 − 2 **1**

6.	7.	8.	9.	10.
5 − 3 **2**	6 − 0 **6**	2 − 2 **0**	4 − 0 **4**	6 − 3 **3**

11.	12.	13.	14.	15.
1 − 1 **0**	5 − 4 **1**	1 − 0 **1**	4 − 1 **3**	3 − 3 **0**

35

Name _____   1.OA.C.6

Subtraction from 20
Solve each problem.

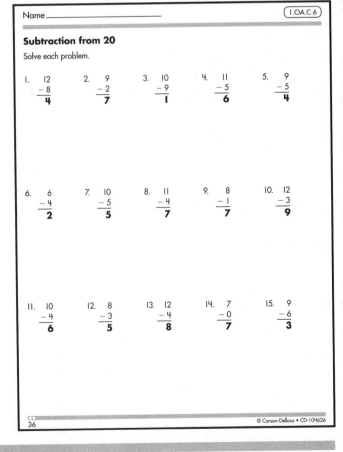

1.	2.	3.	4.	5.
12 − 8 **4**	9 − 2 **7**	10 − 9 **1**	11 − 5 **6**	9 − 5 **4**

6.	7.	8.	9.	10.
6 − 4 **2**	10 − 5 **5**	11 − 4 **7**	8 − 1 **7**	12 − 3 **9**

11.	12.	13.	14.	15.
10 − 4 **6**	8 − 3 **5**	12 − 4 **8**	7 − 0 **7**	9 − 6 **3**

36

111

Answer Key

Subtraction from 20

Solve each problem.

1. $\begin{array}{r} 10 \\ -\ 6 \\ \hline 4 \end{array}$	2. $\begin{array}{r} 12 \\ -\ 7 \\ \hline 5 \end{array}$	3. $\begin{array}{r} 13 \\ -\ 4 \\ \hline 9 \end{array}$	4. $\begin{array}{r} 14 \\ -\ 4 \\ \hline 10 \end{array}$	5. $\begin{array}{r} 20 \\ -\ 8 \\ \hline 12 \end{array}$
6. $\begin{array}{r} 12 \\ -\ 5 \\ \hline 7 \end{array}$	7. $\begin{array}{r} 16 \\ -\ 5 \\ \hline 11 \end{array}$	8. $\begin{array}{r} 18 \\ -\ 9 \\ \hline 9 \end{array}$	9. $\begin{array}{r} 10 \\ -\ 2 \\ \hline 8 \end{array}$	10. $\begin{array}{r} 17 \\ -\ 3 \\ \hline 14 \end{array}$
11. $\begin{array}{r} 11 \\ -\ 3 \\ \hline 8 \end{array}$	12. $\begin{array}{r} 20 \\ -\ 7 \\ \hline 13 \end{array}$	13. $\begin{array}{r} 15 \\ -\ 7 \\ \hline 8 \end{array}$	14. $\begin{array}{r} 19 \\ -\ 9 \\ \hline 10 \end{array}$	15. $\begin{array}{r} 18 \\ -\ 3 \\ \hline 15 \end{array}$

The Equal Sign

Color the two bugs in each box with equal answers. Then, write a number sentence to show they are equal.

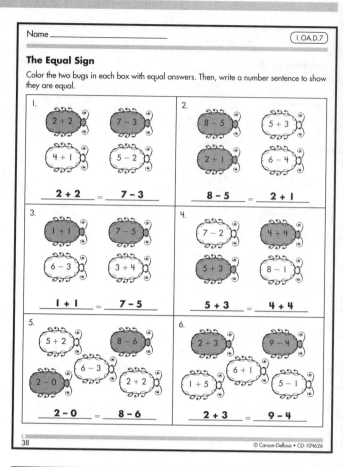

1. $2 + 2$ __ = __ $7 - 3$

2. $8 - 5$ __ = __ $2 + 1$

3. $1 + 1$ __ = __ $7 - 5$

4. $5 + 3$ __ = __ $4 + 4$

5. $2 - 0$ __ = __ $8 - 6$

6. $2 + 3$ __ = __ $9 - 4$

The Equal Sign

Cut out all of the boxes on the right side of the page. Then, glue the boxes in the spaces to make true number sentences.

Order will vary for answers.

1. $\boxed{4 + 4}$ = $\boxed{10 - 2}$

2. $\boxed{3 + 3}$ = $\boxed{7 - 1}$

3. $\boxed{2 + 1}$ = $\boxed{6 - 3}$

4. $\boxed{3 + 2}$ = $\boxed{7 - 2}$

5. $\boxed{6 + 1}$ = $\boxed{9 - 2}$

6. $\boxed{2 + 2}$ = $\boxed{5 - 1}$

The Equal Sign

Find two sets of numbers around the square that are equal. Then, connect their dots with a straight line. Write the number pairs together in a complete number sentence on a line below. Repeat until all of the equal pairs of numbers have been found.

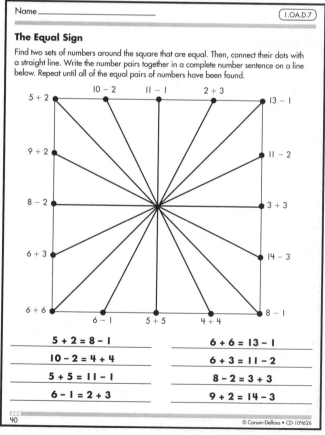

$5 + 2 = 8 - 1$

$10 - 2 = 4 + 4$

$5 + 5 = 11 - 1$

$6 - 1 = 2 + 3$

$6 + 6 = 13 - 1$

$6 + 3 = 11 - 2$

$8 - 2 = 3 + 3$

$9 + 2 = 14 - 3$

Answer Key

True Number Sentences

Look at the numbers on the ends of each balance. Should they be balanced? Circle yes or no.

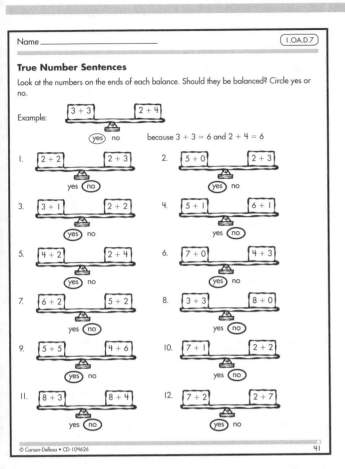

Example:
3 + 3   2 + 4
(yes) no   because 3 + 3 = 6 and 2 + 4 = 6

1. 2 + 2   2 + 3   yes (no)
2. 5 + 0   2 + 3   (yes) no
3. 3 + 1   2 + 2   (yes) no
4. 5 + 1   6 + 1   yes (no)
5. 4 + 2   2 + 4   (yes) no
6. 7 + 0   4 + 3   (yes) no
7. 6 + 2   5 + 2   yes (no)
8. 3 + 3   8 + 0   yes (no)
9. 5 + 5   4 + 6   (yes) no
10. 7 + 1   2 + 2   yes (no)
11. 8 + 3   8 + 4   yes (no)
12. 7 + 2   2 + 7   (yes) no

© Carson-Dellosa • CD-104626   41

True Number Sentences

Circle all of the number sentences below that are true. Then, answer the questions.

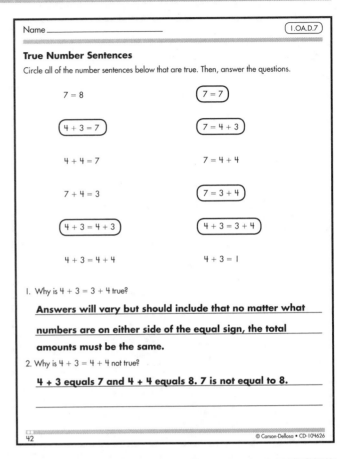

7 = 8   (7 = 7)

(4 + 3 = 7)   (7 = 4 + 3)

4 + 4 = 7   7 = 4 + 4

7 + 4 = 3   (7 = 3 + 4)

(4 + 3 = 4 + 3)   (4 + 3 = 3 + 4)

4 + 3 = 4 + 4   4 + 3 = 1

1. Why is 4 + 3 = 3 + 4 true?

Answers will vary but should include that no matter what numbers are on either side of the equal sign, the total amounts must be the same.

2. Why is 4 + 3 = 4 + 4 not true?

4 + 3 equals 7 and 4 + 4 equals 8. 7 is not equal to 8.

42   © Carson-Dellosa • CD-104626

True Number Sentences

Write numbers in the blanks to make these number sentences true.

1. **3** = 3
2. **4** = 4
3. **5** = 5
4. **9** = 9
5. 7 = **7**
6. 17 = **17**
7. 10 = **10**
8. 88 = **88**
9. **3** + **3** = 6
10. 8 = **5** + **3**

Answers will vary for problems 9–22.

11. **2** + **2** = 4
12. 5 = **1** + **4**
13. **2** + 1 = **3**
14. **10** = 6 + **4**
15. **1** + 3 = **4**
16. **8** = 2 + **6**
17. 5 + **2** = **7**
18. **11** = **8** + 3
19. 4 + **1** = **5**
20. **2** = **1** + 1

21. **2** + **3** + **2** = 7
22. 9 = **1** + **8** + **0**

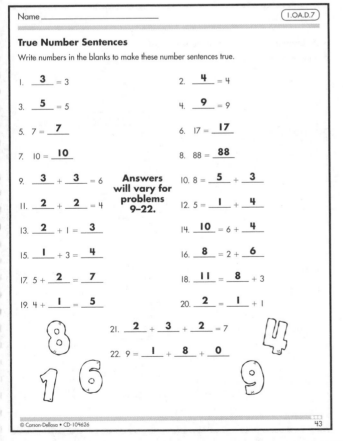

© Carson-Dellosa • CD-104626   43

Addition with an Unknown Number

Draw the missing pictures. Finish the number sentences.

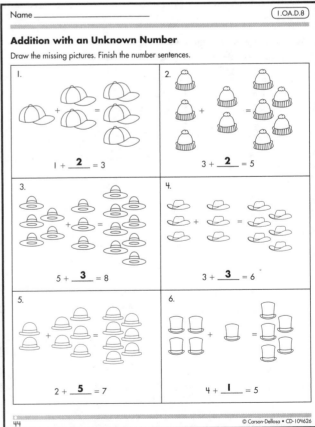

1. 1 + **2** = 3
2. 3 + **2** = 5
3. 5 + **3** = 8
4. 3 + **3** = 6
5. 2 + **5** = 7
6. 4 + **1** = 5

44   © Carson-Dellosa • CD-104626

Answer Key

1.OA.D.8

Addition with an Unknown Number

Draw a picture to solve each problem. Write the answer in the blank.

Example: $4 + \underline{2} = 6$

1. $5 = 2 + \underline{3}$

2. $3 + \underline{6} = 9$

3. $8 = 6 + \underline{2}$

4. $\underline{0} + 4 = 4$

5. $7 = \underline{6} + 1$

6. $\underline{8} + 2 = 10$

7. $8 = \underline{4} + 4$

8. $6 + 3 = \underline{9}$

9. $\underline{10} = 5 + 5$

10. $11 + \underline{4} = 15$

45

1.OA.D.8

Addition with an Unknown Number

The Math Whiz is great at math, but she's not perfect. See if you can find her mistakes. Draw a picture to solve the problem. Write the correct answer in the blank to complete the number sentence. Tell whether she is right.

1. $6 + \underline{4} = 10$

   The Math Whiz says 4.
   Is she right? __yes__

2. $7 = 6 + \underline{1}$

   The Math Whiz says 1.
   Is she right? __yes__

3. $7 - \underline{5} = 2$

   The Math Whiz says 5.
   Is she right? __yes__

4. $3 = 4 - \underline{1}$

   The Math Whiz says 7.
   Is she right? __no__

5. $\underline{5} + 4 = 9$

   The Math Whiz says 5.
   Is she right? __yes__

6. $10 = \underline{2} + 8$

   The Math Whiz says 18.
   Is she right? __no__

7. $\underline{6} - 4 = 2$

   The Math Whiz says 2.
   Is she right? __no__

8. $1 = \underline{6} - 5$

   The Math Whiz says 6.
   Is she right? __yes__

1.OA.D.8

Subtraction with an Unknown Number

Draw a picture to solve each problem. Write the answer in the blank.

Example: $9 - 6 = \underline{}$

1. $\underline{4} = 6 - 2$

2. $7 - 4 = \underline{3}$

3. $\underline{2} = 9 - 7$

4. $5 - \underline{3} = 2$

5. $1 = 4 - \underline{3}$

6. $6 - \underline{0} = 6$

7. $7 = 8 - \underline{1}$

8. $\underline{6} - 3 = 3$

9. $1 = \underline{5} - 4$

10. $12 - \underline{3} = 9$

47

1.OA.D.8

Subtraction with an Unknown Number

Pluswoman and Minusman have both given an answer to each problem below. Circle the answer that is correct.

1. $\underline{5} - 3 = 2$

   Pluswoman says 4.
   (Minusman says 5.)

2. $4 = \underline{5} - 1$

   (Pluswoman says 5.)
   Minusman says 3.

3. $\underline{7} - 2 = 5$

   (Pluswoman says 7.)
   Minusman says 3.

4. $2 = \underline{6} - 4$

   Pluswoman says 2.
   (Minusman says 6.)

5. $\underline{8} - 1 = 7$

   Pluswoman says 6.
   (Minusman says 8.)

6. $6 = \underline{8} - 2$

   (Pluswoman says 8.)
   Minusman says 4.

7. $\underline{8} - 3 = 5$

   Pluswoman says 2.
   (Minusman says 8.)

8. $4 = \underline{7} - 3$

   Pluswoman says 1.
   (Minusman says 7.)

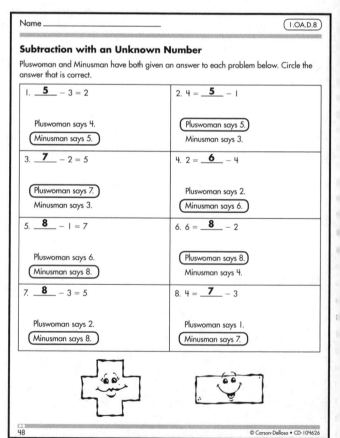

Answer Key

Subtraction with an Unknown Number

Find the fish with the correct answer and shade it. Write the answer in the blank to complete each number sentence.

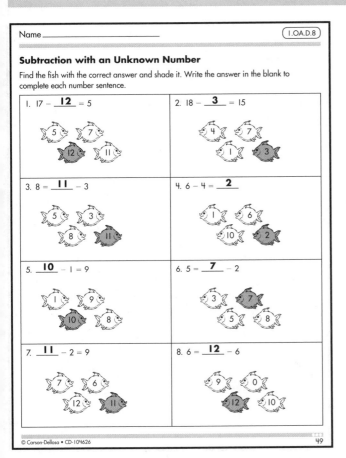

1. $17 - \underline{\textbf{12}} = 5$
2. $18 - \underline{\textbf{3}} = 15$
3. $8 = \underline{\textbf{11}} - 3$
4. $6 - 4 = \underline{\textbf{2}}$
5. $\underline{\textbf{10}} - 1 = 9$
6. $5 = \underline{\textbf{7}} - 2$
7. $\underline{\textbf{11}} - 2 = 9$
8. $6 = \underline{\textbf{12}} - 6$

Counting to 120

Write the correct number in each box.

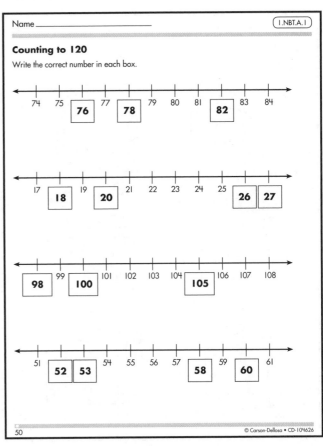

74   75   **76**   77   **78**   79   80   81   **82**   83   84

17   **18**   19   **20**   21   22   23   24   25   **26**   **27**

98   99   **100**   101   102   103   104   **105**   106   107   108

51   **52**   **53**   54   55   56   57   **58**   59   **60**   61

Counting to 120

Write each missing number.

1	2	**3**	**4**	**5**	6	**7**	8	9	**10**
11	**12**	**13**	**14**	**15**	16	**17**	**18**	**19**	**20**
21	**22**	**23**	**24**	25	**26**	27	28	**29**	30
31	32	**33**	**34**	35	**36**	**37**	**38**	39	**40**
41	**42**	43	**44**	**45**	46	47	**48**	**49**	50
51	52	**53**	54	55	**56**	**57**	58	**59**	**60**
61	**62**	**63**	64	**65**	**66**	67	68	**69**	**70**
71	**72**	73	**74**	75	**76**	**77**	**78**	79	80
81	82	**83**	**84**	**85**	**86**	87	**88**	**89**	**90**
91	**92**	93	**94**	**95**	96	**97**	**98**	**99**	100
101	102	103	**104**	105	**106**	107	108	**109**	110
111	**112**	**113**	114	**115**	**116**	**117**	118	119	**120**

Counting to 120

Write the number that comes one before.

1. **4**   5
2. **88**   89
3. **22**   23
4. **116**   117

Write the number that comes between.

5. 11   **12**   13
6. 35   **36**   37
7. 62   **63**   64
8. 98   **99**   100

Write the number that comes one after.

9. 79   **80**
10. 92   **93**
11. 50   **51**
12. 114   **115**
13. 41   **42**
14. 103   **104**

Answer Key

Name _____

Tens and Ones

Circle the correct number of blocks to match the number on each gift.

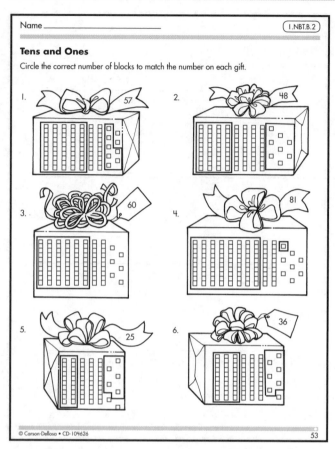

53

Name _____

Tens and Ones

Draw lines to match the equal numbers.

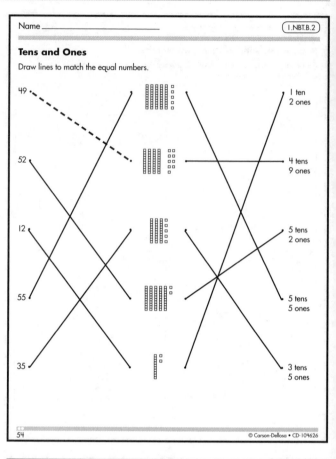

54

Name _____

Tens and Ones

Write the number for each group of blocks. Then, use the code to answer the riddle.

Why does a math teacher comb her hair?

To get out the R E C T A N G L E S !
              37 64 73 45 20 35 28 54 64 15

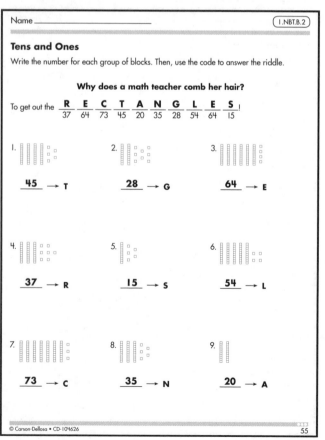

1. **45** → T
2. **28** → G
3. **64** → E
4. **37** → R
5. **15** → S
6. **54** → L
7. **73** → C
8. **35** → N
9. **20** → A

55

Name _____

Place Value

Write how many tens and ones are in each number.

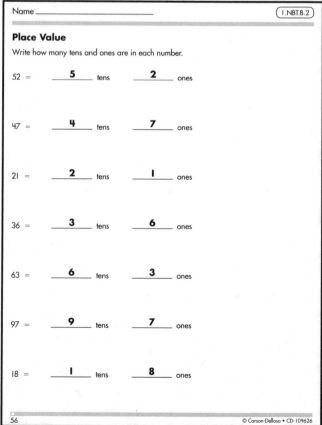

52 = **5** tens **2** ones

47 = **4** tens **7** ones

21 = **2** tens **1** ones

36 = **3** tens **6** ones

63 = **6** tens **3** ones

97 = **9** tens **7** ones

18 = **1** tens **8** ones

56

Answer Key

Name _____ 1.NBT.B.2

Place Value

Write each number.

1. 5 tens   4 ones = **54**   2. 2 tens   7 ones = **27**

3. 8 tens   9 ones = **89**   4. 7 tens   5 ones = **75**

5. 1 ten   6 ones = **16**   6. 4 ten   3 ones = **43**

7. 6 tens   0 ones = **60**   8. 9 tens   1 one = **91**

9. 0 tens   8 ones = **8**   10. 3 tens   2 ones = **32**

Write the number of tens and ones for each number.

11. 71 = **7** tens   **1** one

12. 58 = **5** tens   **8** ones

13. 5 = **0** tens   **5** ones

14. 40 = **4** tens   **0** ones

57

Name _____ 1.NBT.B.2

Place Value

Write each number in three different ways.

1. 74 = **7** tens   **4** ones
   74 = **6** tens   **14** ones
   74 = **5** tens   **24** ones

2. 88 = **8** tens   **8** ones
   88 = **7** tens   **18** ones
   88 = **6** tens   **28** ones

3. 46 = **3** tens   **6** ones
   46 = **2** tens   **16** ones
   46 = **1** tens   **26** ones

4. 63 = **6** tens   **3** ones
   63 = **5** tens   **13** ones
   63 = **4** tens   **23** ones

58

Name _____ 1.NBT.B.2

Understanding Two-Digit Numbers

Use the code to color the jar of jelly beans.

Color the jelly beans with 0 ones orange.
Color the jelly beans with 2 tens green.
Color the jelly beans with 3 tens red.
Color the jelly beans with 7 tens yellow.
Color the jelly beans with 9 ones purple.
Color the jelly beans with 6 ones blue.

59

Name _____ 1.NBT.B.2

Understanding Two-Digit Numbers

Write the numbers on the left in the correct sets.

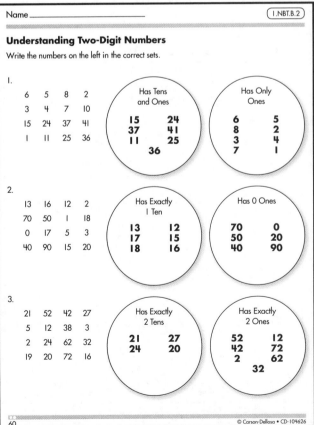

1.
6	5	8	2
3	4	7	10
15	24	37	41
1	11	25	36

Has Tens and Ones: **15 24 37 41 11 25 36**

Has Only Ones: **6 5 8 2 3 4 7 1**

2.
13	16	12	2
70	50	1	18
0	17	5	3
40	90	15	20

Has Exactly 1 Ten: **13 12 17 15 18 16**

Has 0 Ones: **70 0 50 20 40 90**

3.
21	52	42	27
5	12	38	3
2	24	62	32
19	20	72	16

Has Exactly 2 Tens: **21 27 24 20**

Has Exactly 2 Ones: **52 12 42 72 2 62 32**

60

Answer Key

Understanding Two-Digit Numbers

Some numbers have been sorted into sets. Cut out and glue each set name under the correct set.

1. **less than 3 tens**

2. **more than 5 tens**

3. **3 in the tens place**

4. **5 in the ones place**

Comparing Numbers

Follow the directions for each set of numbers.

1. Draw a triangle around the number that is equal to 7.
   Circle the numbers that are less than 7.
   Draw boxes around the numbers that are greater than 7.

2. Draw a triangle around the number that is equal to 25.
   Circle the numbers that are less than 25.
   Draw boxes around the numbers that are greater than 25.

3. Draw a triangle around the number that is equal to 30.
   Circle the numbers that are less than 30.
   Draw boxes around the numbers that are greater than 30.

4. Draw a triangle around the number that is equal to 46.
   Circle the numbers that are less than 46.
   Draw boxes around the numbers that are greater than 46.

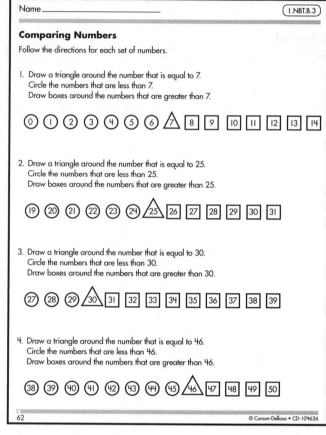

Comparing Numbers

Write the number in each box that answers the riddle.

1. (4) (6) (8) (12)

   I am less than 10.
   I am greater than 5.
   I am not equal to 8.

   What number am I? **6**

2. (12) (14) (17) (9)

   I am less than 15.
   I am greater than 10.
   I am not equal to 12.

   What number am I? **14**

3. (5) (2) (4) (7)

   I am less than 6.
   I am greater than 3.
   I am not equal to 4.

   What number am I? **5**

4. (25) (16) (15) (10)

   I am less than 20.
   I am greater than 13.
   I am not equal to 15.

   What number am I? **16**

5. (30) (26) (23) (27)

   I am less than 30.
   I am greater than 24.
   I am not equal to 26.

   What number am I? **27**

6. (35) (50) (32) (29)

   I am less than 40.
   I am greater than 30.
   I am not equal to 35.

   What number am I? **32**

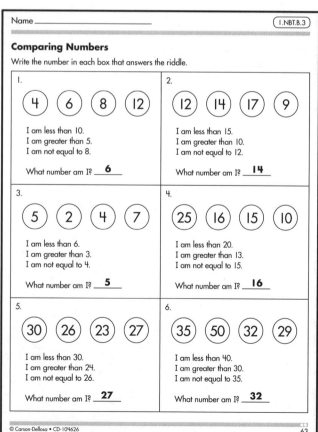

Comparing Numbers

Write the numbers in order from least to greatest on the mailboxes.

1. 53   27   19   **19**   **27**   **53**

2. 48   31   64   **31**   **48**   **64**

3. 12   36   25   **12**   **25**   **36**

4. 83   74   67   **67**   **74**   **83**

Use <, >, or = to compare the numbers. Use the numbers above to help you.

5. 19 (<) 27   6. 48 (>) 31

7. 12 (<) 36   8. 83 (=) 83

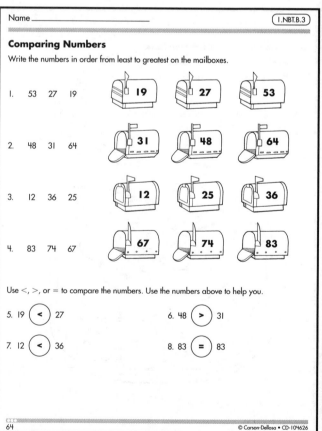

Answer Key

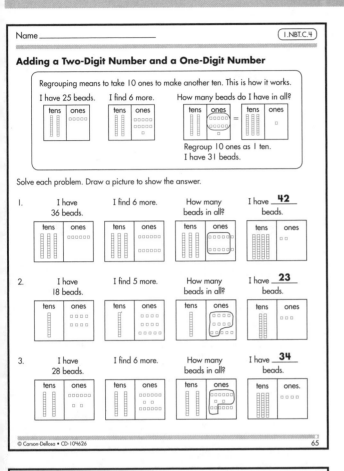

Name _____ 1.NBT.C.4

Adding a Two-Digit Number and a One-Digit Number

Regrouping means to take 10 ones to make another ten. This is how it works.

I have 25 beads. I find 6 more. How many beads do I have in all?

Regroup 10 ones as 1 ten.
I have 31 beads.

Solve each problem. Draw a picture to show the answer.

1. I have 36 beads. I find 6 more. How many beads in all? I have **42** beads.

2. I have 18 beads. I find 5 more. How many beads in all? I have **23** beads.

3. I have 28 beads. I find 6 more. How many beads in all? I have **34** beads.

© Carson-Dellosa • CD-104626   65

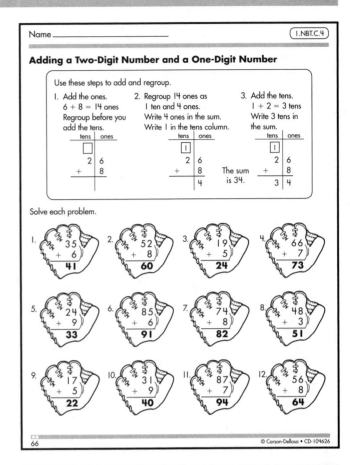

Name _____ 1.NBT.C.4

Adding a Two-Digit Number and a One-Digit Number

Use these steps to add and regroup.

1. Add the ones.
   6 + 8 = 14 ones
   Regroup before you add the tens.

2. Regroup 14 ones as 1 ten and 4 ones.
   Write 4 ones in the sum.
   Write 1 in the tens column.

3. Add the tens.
   1 + 2 = 3 tens
   Write 3 tens in the sum.
   The sum is 34.

Solve each problem.

1. 35 + 6 = **41**
2. 52 + 8 = **60**
3. 19 + 5 = **24**
4. 66 + 7 = **73**
5. 24 + 9 = **33**
6. 85 + 6 = **91**
7. 74 + 8 = **82**
8. 48 + 3 = **51**
9. 17 + 5 = **22**
10. 31 + 9 = **40**
11. 87 + 7 = **94**
12. 56 + 8 = **64**

66   © Carson-Dellosa • CD-104626

Name _____ 1.NBT.C.4

Adding a Two-Digit Number and a One-Digit Number

Look at this!
```
  tens | ones
    1
    7  |  3
 +     |  9
    8  |  2
```

Solve the problems in order. Cross out each sum on the frozen ice-cream treats to see which treat is finished first.

1. 56 + 8 = **64**
2. 59 + 9 = **68**
3. 29 + 4 = **33**
4. 79 + 2 = **81**
5. 35 + 6 = **41**

6. 47 + 5 = **52**
7. 12 + 8 = **20**
8. 72 + 9 = **81**
9. 57 + 7 = **64**

10. 44 + 6 = **50**
11. 49 + 6 = **55**
12. 63 + 8 = **71**

13. 54 + 7 = **61**
11. 25 + 5 = **30**
12. 82 + 8 = **90**

© Carson-Dellosa • CD-104626   67

Name _____ 1.NBT.C.4

Adding Multiples of 10

Add a ten block to each group. Write the number sentence.

1. **44** + **10** = **54**
2. **61** + **10** = **71**
3. **73** + **10** = **83**
4. **18** + **10** = **28**
5. **29** + **10** = **39**
6. **55** + **10** = **65**

Complete the table by adding 10 to each number on the left.

+ 10	
62	**72**
33	**43**
87	**97**
5	**15**
56	**66**
21	**31**

68   © Carson-Dellosa • CD-104626

© Carson-Dellosa • CD-104626

119

Answer Key

Adding Multiples of 10

Which bowling pin will fall next? To find out, solve the problems in order. Cross off each sum on the bowling pins. The bowling pin with all of the numbers crossed off is the next one to fall. Color it.

1. 43
    + 10
    53

2. 16
    + 10
    26

3. 71
    + 10
    81

4. 24
    + 10
    34

5. 85
    + 10
    95

6. 39
    + 10
    49

7. 15
    + 10
    25

8. 46
    + 10
    56

9. 57
    + 10
    67

10. 63
    + 10
    73

11. 24
    + 10
    34

12. 73
    + 10
    83

58 26 34 28 67 34 83

84 95 49 56 73 86 96

© Carson-Dellosa • CD-104626   69

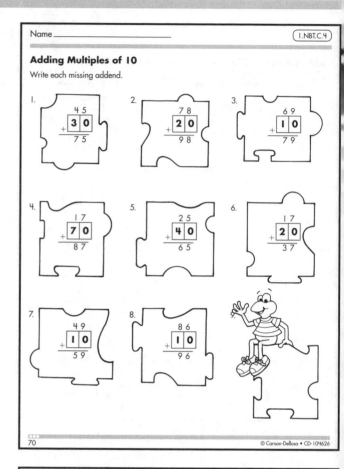

Adding Multiples of 10

Write each missing addend.

1. 4 5
    + **3 0**
    7 5

2. 7 8
    + **2 0**
    9 8

3. 6 9
    + **1 0**
    7 9

4. 1 7
    + **7 0**
    8 7

5. 2 5
    + **4 0**
    6 5

6. 1 7
    + **2 0**
    3 7

7. 4 9
    + **1 0**
    5 9

8. 8 6
    + **1 0**
    9 6

70   © Carson-Dellosa • CD-104626

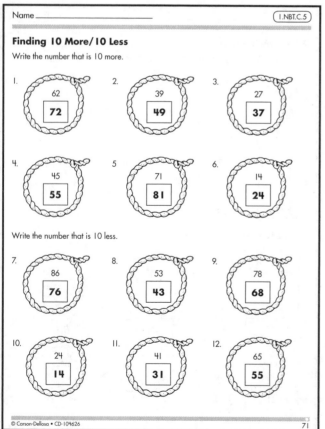

Finding 10 More/10 Less

Write the number that is 10 more.

1. 62
    72

2. 39
    49

3. 27
    37

4. 45
    55

5. 71
    81

6. 14
    24

Write the number that is 10 less.

7. 86
    76

8. 53
    43

9. 78
    68

10. 24
    14

11. 41
    31

12. 65
    55

© Carson-Dellosa • CD-104626   71

Finding 10 More/10 Less

Look at each number. Write the number that is 10 more.

53 **63**   49 **59**   21 **31**   70 **80**

34 **44**   72 **82**   38 **48**   16 **26**   25 **35**

Look at each number. Write the number that is 10 less.

83 **73**   69 **59**   57 **47**   90 **80**   34 **24**

16 **6**   22 **12**   63 **53**   85 **75**   Bank

72   © Carson-Dellosa • CD-104626

Answer Key

(1.NBT.C.5)

Finding 10 More/10 Less

Find two numbers around the square that have a difference of 10. Connect their dots with a straight line. Repeat until each dot has been connected to another.

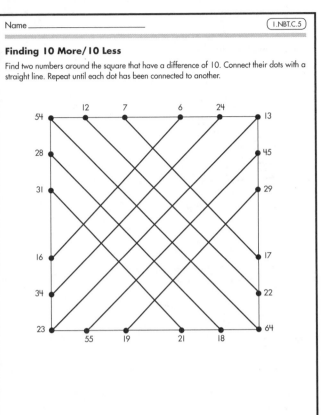

(1.NBT.C.6)

Subtracting Multiples of 10

Take 10 seeds away from each picture. Write the number.

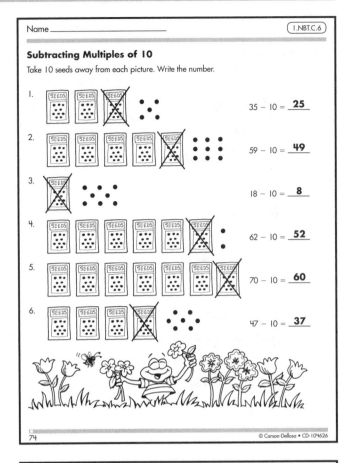

1. $35 - 10 = $ **25**

2. $59 - 10 = $ **49**

3. $18 - 10 = $ **8**

4. $62 - 10 = $ **52**

5. $70 - 10 = $ **60**

6. $47 - 10 = $ **37**

(1.NBT.C.6)

Subtracting Multiples of 10

Solve each problem.

1. $\begin{array}{r} 63 \\ -10 \\ \hline \mathbf{53} \end{array}$	2. $\begin{array}{r} 80 \\ -10 \\ \hline \mathbf{70} \end{array}$	3. $\begin{array}{r} 75 \\ -10 \\ \hline \mathbf{65} \end{array}$	4. $\begin{array}{r} 79 \\ -10 \\ \hline \mathbf{69} \end{array}$	5. $\begin{array}{r} 38 \\ -10 \\ \hline \mathbf{28} \end{array}$	6. $\begin{array}{r} 93 \\ -10 \\ \hline \mathbf{83} \end{array}$
7. $\begin{array}{r} 67 \\ -10 \\ \hline \mathbf{57} \end{array}$	8. $\begin{array}{r} 83 \\ -10 \\ \hline \mathbf{73} \end{array}$	9. $\begin{array}{r} 77 \\ -10 \\ \hline \mathbf{67} \end{array}$	10. $\begin{array}{r} 76 \\ -10 \\ \hline \mathbf{66} \end{array}$	11. $\begin{array}{r} 59 \\ -10 \\ \hline \mathbf{49} \end{array}$	12. $\begin{array}{r} 57 \\ -10 \\ \hline \mathbf{47} \end{array}$
13. $\begin{array}{r} 56 \\ -10 \\ \hline \mathbf{46} \end{array}$	14. $\begin{array}{r} 63 \\ -10 \\ \hline \mathbf{53} \end{array}$	15. $\begin{array}{r} 48 \\ -10 \\ \hline \mathbf{38} \end{array}$	16. $\begin{array}{r} 86 \\ -10 \\ \hline \mathbf{76} \end{array}$	17. $\begin{array}{r} 40 \\ -10 \\ \hline \mathbf{30} \end{array}$	18. $\begin{array}{r} 43 \\ -10 \\ \hline \mathbf{33} \end{array}$

(1.NBT.C.6)

Subtracting Multiples of 10

Solve each problem.

1. $\begin{array}{r} 77 \\ -40 \\ \hline \mathbf{37} \end{array}$	2. $\begin{array}{r} 55 \\ -20 \\ \hline \mathbf{35} \end{array}$	3. $\begin{array}{r} 98 \\ -50 \\ \hline \mathbf{48} \end{array}$	4. $\begin{array}{r} 29 \\ -10 \\ \hline \mathbf{19} \end{array}$	5. $\begin{array}{r} 60 \\ -50 \\ \hline \mathbf{10} \end{array}$	6. $\begin{array}{r} 69 \\ -20 \\ \hline \mathbf{49} \end{array}$
7. $\begin{array}{r} 45 \\ -20 \\ \hline \mathbf{25} \end{array}$	8. $\begin{array}{r} 78 \\ -60 \\ \hline \mathbf{18} \end{array}$	9. $\begin{array}{r} 86 \\ -80 \\ \hline \mathbf{6} \end{array}$	10. $\begin{array}{r} 39 \\ -10 \\ \hline \mathbf{29} \end{array}$	11. $\begin{array}{r} 86 \\ -20 \\ \hline \mathbf{66} \end{array}$	12. $\begin{array}{r} 59 \\ -50 \\ \hline \mathbf{9} \end{array}$
13. $\begin{array}{r} 72 \\ -30 \\ \hline \mathbf{42} \end{array}$	14. $\begin{array}{r} 93 \\ -80 \\ \hline \mathbf{13} \end{array}$	15. $\begin{array}{r} 26 \\ -10 \\ \hline \mathbf{16} \end{array}$	16. $\begin{array}{r} 98 \\ -10 \\ \hline \mathbf{88} \end{array}$	17. $\begin{array}{r} 67 \\ -30 \\ \hline \mathbf{37} \end{array}$	18. $\begin{array}{r} 39 \\ -10 \\ \hline \mathbf{29} \end{array}$
19. $\begin{array}{r} 77 \\ -10 \\ \hline \mathbf{67} \end{array}$	20. $\begin{array}{r} 87 \\ -10 \\ \hline \mathbf{77} \end{array}$	21. $\begin{array}{r} 63 \\ -30 \\ \hline \mathbf{33} \end{array}$	22. $\begin{array}{r} 82 \\ -70 \\ \hline \mathbf{12} \end{array}$	23. $\begin{array}{r} 74 \\ -50 \\ \hline \mathbf{24} \end{array}$	24. $\begin{array}{r} 83 \\ -40 \\ \hline \mathbf{43} \end{array}$

Answer Key

Comparing Lengths

Color the longest object in each set red. Color the shortest object in each set blue.

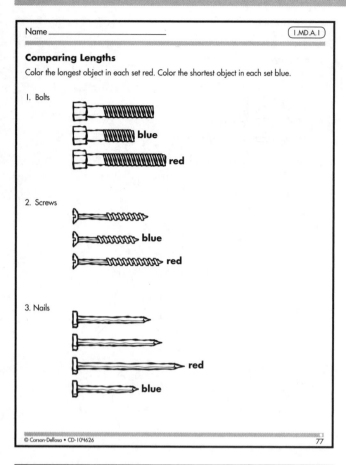

1. Bolts

   blue
   red

2. Screws

   blue
   red

3. Nails

   red
   blue

Comparing Lengths

Each clown's shoe is a different length. Write the clowns' names in order of the lengths of their shoes from shortest to longest.

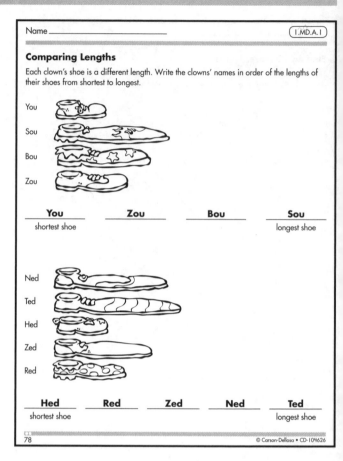

You

Sou

Bou

Zou

You	**Zou**	**Bou**	**Sou**
shortest shoe			longest shoe

Ned

Ted

Hed

Zed

Red

Hed	**Red**	**Zed**	**Ned**	**Ted**
shortest shoe				longest shoe

Comparing Lengths

Each clown's hat has a letter. Look at the heights of the hats and answer the questions using the letters for the hats.

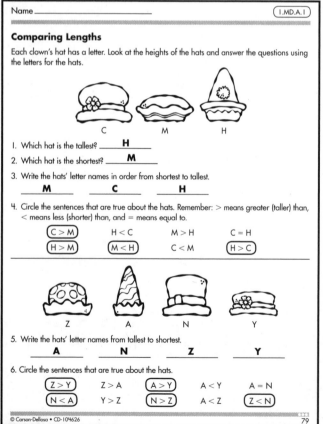

C   M   H

1. Which hat is the tallest? **H**

2. Which hat is the shortest? **M**

3. Write the hats' letter names in order from shortest to tallest.
   M   **C**   **H**

4. Circle the sentences that are true about the hats. Remember: > means greater (taller) than, < means less (shorter) than, and = means equal to.

   (C > M)   H < C   M > H   C = H
   (H > M)   (M < H)   C < M   (H > C)

   Z   A   N   Y

5. Write the hats' letter names from tallest to shortest.
   A   **N**   **Z**   **Y**

6. Circle the sentences that are true about the hats.

   (Z > Y)   Z > A   (A > Y)   A < Y   A = N
   (N < A)   Y > Z   (N > Z)   A < Z   (Z < N)

Measuring in Units

Write how long each object is in units.

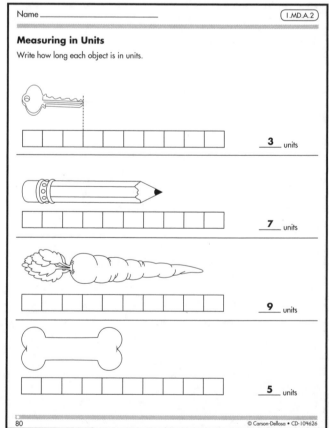

3 units

7 units

9 units

5 units

Answer Key

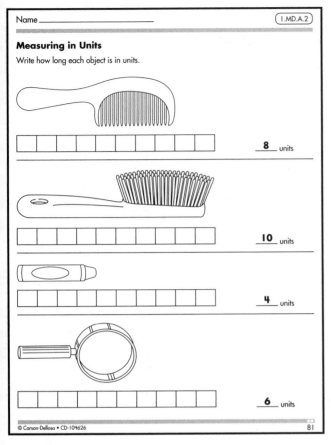

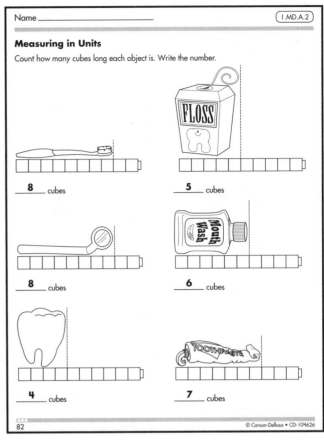

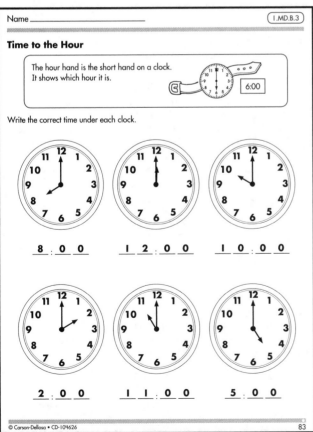

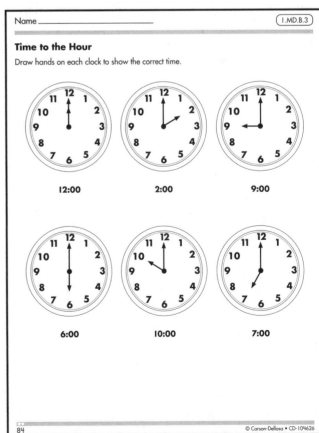

Answer Key

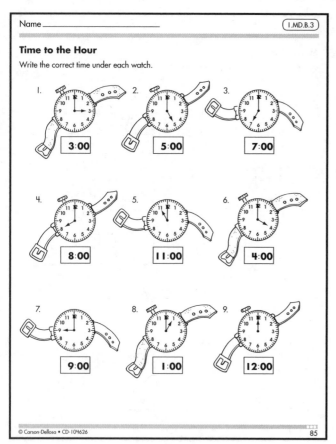

Time to the Hour

Write the correct time under each watch.

1. 3:00
2. 5:00
3. 7:00
4. 8:00
5. 11:00
6. 4:00
7. 9:00
8. 1:00
9. 12:00

85

Time to the Half Hour

The minute hand is the longer hand. It is pointing to the 6. Thirty minutes have passed after 3:00. The time is 3:30.

Write the correct time under each clock.

7 : 3 0   6 : 3 0   1 2 : 3 0

9 : 3 0   2 : 3 0   4 : 3 0

86

Time to the Half Hour

Draw hands on each clock to show the correct time.

4:30   6:30   1:30

10:30   9:30   12:30

87

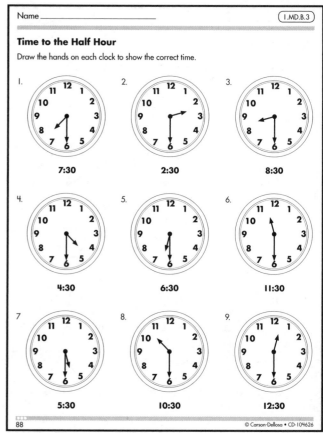

Time to the Half Hour

Draw the hands on each clock to show the correct time.

1. 7:30
2. 2:30
3. 8:30
4. 4:30
5. 6:30
6. 11:30
7. 5:30
8. 10:30
9. 12:30

88

Answer Key

Analyzing Data

Ms. Smith's first grade class made a tally chart about their pets. Each student made only one tally. Use the chart to answer the questions.

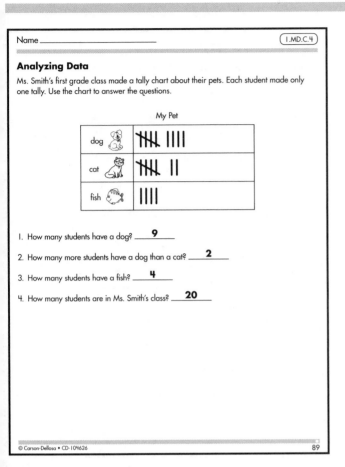

My Pet

dog	$\cancel{				}\				$
cat	$\cancel{				}\		$
fish	$				$

1. How many students have a dog? __**9**__

2. How many more students have a dog than a cat? __**2**__

3. How many students have a fish? __**4**__

4. How many students are in Ms. Smith's class? __**20**__

Analyzing Data

Mrs. Mason's class has earned an extra recess! They voted to decide what to do. Use the tally chart to answer the question.

Extra Recess Games

Kickball	$\cancel{				}\			$
Four square	$\cancel{				}$
Board games	$			$

1. Which game was chosen the least? __**board games**__

2. How many students are in Mrs. Mason's class? __**16**__

3. Which game was chosen the most? __**kickball**__

4. How many more votes did four square get than board games? __**2**__

Analyzing Data

Ask 10 people which drink they like best. In the table below, make a tally mark beside the drink each one likes. Use the chart to answer the questions.

Favorite Kind of Drink

juice	
milk	
water	

1. How many people like milk? _____

2. How many people like juice? _____   **Answers will vary.**

3. Do more people like water or milk? _____

4. How many people like water and juice in all? _____

Attributes of Two-Dimensional Shapes

Cut out the shapes below. Glue each shape to its matching shape. Write the number of sides and corners.

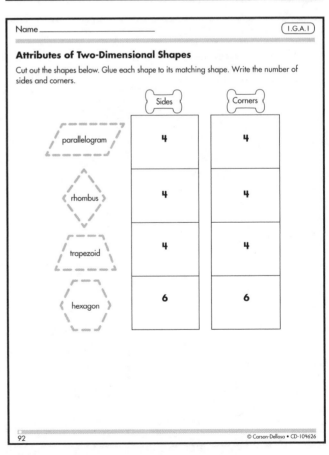

	Sides	Corners
parallelogram	4	4
rhombus	4	4
trapezoid	4	4
hexagon	6	6

Answer Key

Attributes of Two-Dimensional Shapes

Look at the picture of a log cabin. How many of each shape can you find in the picture?

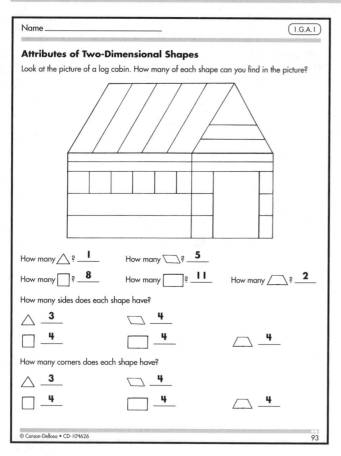

How many △? **1**　　　How many ▱? **5**

How many ☐? **8**　　　How many ▭? **11**　　　How many ▱? **2**

How many sides does each shape have?

△ **3**　　　▱ **4**

☐ **4**　　　▭ **4**　　　▱ **4**

How many corners does each shape have?

△ **3**　　　▱ **4**

☐ **4**　　　▭ **4**　　　▱ **4**

Attributes of Two-Dimensional Shapes

Jan made cookies with her new cookie cutters. Count how many sides each cookie has. Then, answer the questions below.

6 sides　　**4** sides　　**3** sides　　**8** sides　　**5** sides

H　　　　S　　　　T　　　　O　　　　P

hexagon cookie　　square cookie　　triangle cookie　　octagon cookie　　pentagon cookie

1. Which cookie has the fewest sides? **triangle cookie**

2. Which cookie has the most sides? **octagon cookie**

3. Write the cookie letters in order of the number of sides the shapes have.

T　　**S**　　**P**　　**H**　　**O**
fewest sides　　　　　　　　　　　most sides

4. How many more sides does O have than H? **2**

5. How many more sides does H have than S? **2**

6. How many fewer sides does T have than P? **2**

7. How many fewer sides does S have than O? **4**

8. How many more sides would P need to become O? **3**

Attributes of Three-Dimensional Shapes

The flat side of a solid shape is called a face.　　This is a face.

Complete the table.

Spatial Shape	Number of Faces
△ (cone)	1
☐ (cube)	6
cylinder	2
rectangular prism	6
sphere	0

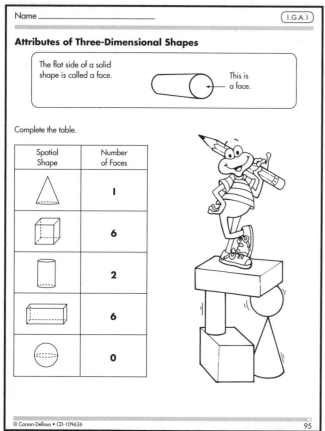

Attributes of Three-Dimensional Shapes

Complete the table.

	Number of Faces	Number of Edges	Number of Vertices (Corners)
cone	1	0	0
cube	6	12	8
cylinder	2	0	0
pyramid	5	8	5

Answer Key

Name _____ I.G.A.1

Attributes of Three-Dimensional Shapes

Cut out the shapes. Glue each shape in the correct box.

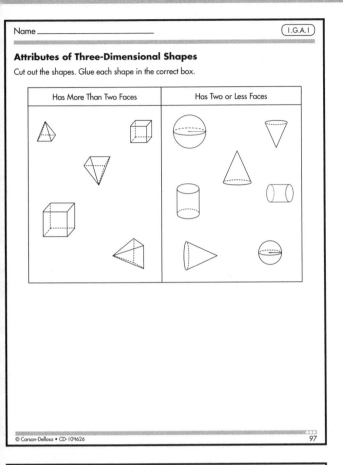

© Carson-Dellosa • CD-104626   97

Name _____ I.G.A.2

Composing Shapes

Look at the figure. Circle the set of shapes that form this figure when put together.

Tell or write why you chose that set of shapes.

Answers will vary.

_____

_____

98   © Carson-Dellosa • CD-104626

Name _____ I.G.A.2

Composing Shapes

Look at the figure. Circle the set of shapes that form this figure when put together.

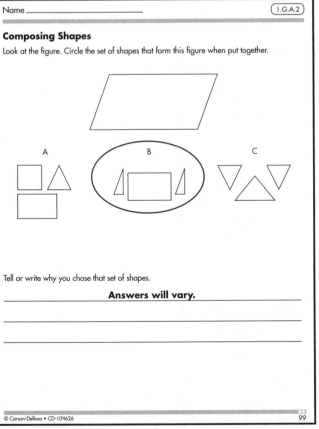

Tell or write why you chose that set of shapes.

Answers will vary.

_____

_____

© Carson-Dellosa • CD-104626   99

Name _____ I.G.A.2

Composing Shapes

Look at the figure. Circle the set of shapes that form this figure when put together.

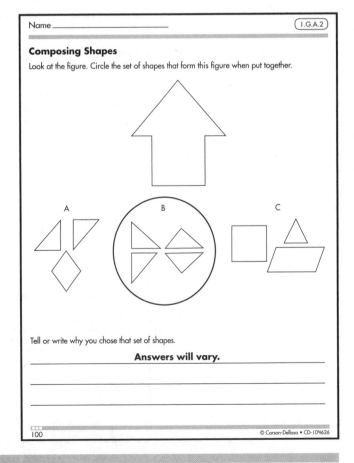

Tell or write why you chose that set of shapes.

Answers will vary.

_____

_____

100   © Carson-Dellosa • CD-104626

Answer Key

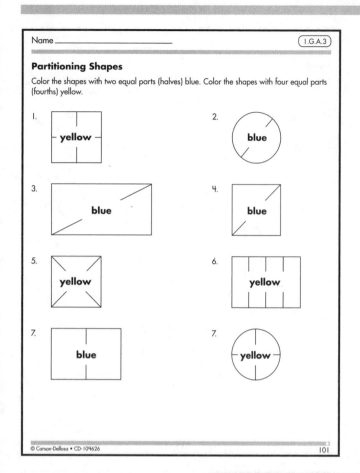

Partitioning Shapes

Color the shapes with two equal parts (halves) blue. Color the shapes with four equal parts (fourths) yellow.

1. yellow
2. blue
3. blue
4. blue
5. yellow
6. yellow
7. blue
7. yellow

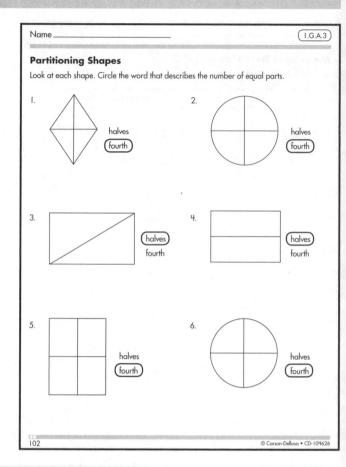

Partitioning Shapes

Look at each shape. Circle the word that describes the number of equal parts.

1. halves / **fourth**
2. halves / **fourth**
3. **halves** / fourth
4. **halves** / fourth
5. halves / **fourth**
6. halves / **fourth**

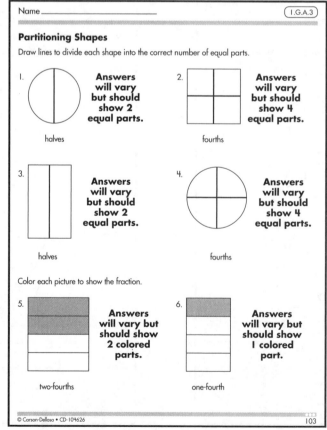

Partitioning Shapes

Draw lines to divide each shape into the correct number of equal parts.

1. **Answers will vary but should show 2 equal parts.**
   halves

2. **Answers will vary but should show 4 equal parts.**
   fourths

3. **Answers will vary but should show 2 equal parts.**
   halves

4. **Answers will vary but should show 4 equal parts.**
   fourths

Color each picture to show the fraction.

5. **Answers will vary but should show 2 colored parts.**
   two-fourths

6. **Answers will vary but should show 1 colored part.**
   one-fourth

Congratulations!

receives this award for

_____

Signed _____

Date _____

1 + 1	1 + 2	1 + 3	1 + 4
1 + 5	1 + 6	1 + 7	1 + 8
1 + 9	2 + 2	2 + 3	2 + 4
2 + 5	2 + 6	2 + 7	2 + 8

5	4	3	2
<u>9</u>	8	7	<u>6</u>
<u>6</u>	5	4	10
10	<u>9</u>	8	7

2 + 9	3 + 3	3 + 4	3 + 5
3 + 6	3 + 7	3 + 8	3 + 9
4 + 4	4 + 5	4 + 6	4 + 7
4 + 8	4 + 9	5 + 5	5 + 6

8 7 <u>6</u> 11

12 11 10 <u>9</u>

11 10 <u>9</u> 8

11 10 13 12

5 + 7	5 + 8	5 + 9	6 + 6
6 + 7	6 + 8	6 + 9	7 + 7
7 + 8	7 + 9	8 + 8	8 + 9
9 + 9	1 − 1	2 − 2	2 − 1

12	14	13	12
14	15	14	13
17	16	16	15
1	0	0	18

3 − 3	3 − 2	3 − 1	4 − 4
© CD	© CD	© CD	© CD
4 − 3	4 − 2	4 − 1	5 − 5
© CD	© CD	© CD	© CD
5 − 4	5 − 3	5 − 2	5 − 1
© CD	© CD	© CD	© CD
6 − 6	6 − 5	6 − 4	6 − 3
© CD	© CD	© CD	© CD

0 2 1 0

0 3 2 1

4 3 2 1

3 2 1 0

$\begin{array}{r} 6 \\ -\ 2 \\ \hline \end{array}$	$\begin{array}{r} 6 \\ -\ 1 \\ \hline \end{array}$	$\begin{array}{r} 7 \\ -\ 7 \\ \hline \end{array}$	$\begin{array}{r} 7 \\ -\ 6 \\ \hline \end{array}$
$\begin{array}{r} 7 \\ -\ 5 \\ \hline \end{array}$	$\begin{array}{r} 7 \\ -\ 4 \\ \hline \end{array}$	$\begin{array}{r} 7 \\ -\ 3 \\ \hline \end{array}$	$\begin{array}{r} 7 \\ -\ 2 \\ \hline \end{array}$
$\begin{array}{r} 7 \\ -\ 1 \\ \hline \end{array}$	$\begin{array}{r} 8 \\ -\ 8 \\ \hline \end{array}$	$\begin{array}{r} 8 \\ -\ 7 \\ \hline \end{array}$	$\begin{array}{r} 8 \\ -\ 6 \\ \hline \end{array}$
$\begin{array}{r} 8 \\ -\ 5 \\ \hline \end{array}$	$\begin{array}{r} 8 \\ -\ 4 \\ \hline \end{array}$	$\begin{array}{r} 8 \\ -\ 3 \\ \hline \end{array}$	$\begin{array}{r} 8 \\ -\ 2 \\ \hline \end{array}$

1	0	5	4
5	4	3	2
2	1	0	<u>6</u>
<u>6</u>	5	4	3

8 − 1	9 − 9	9 − 8	9 − 7
9 − 6	9 − 5	9 − 4	9 − 3
9 − 2	9 − 1	10 − 0	10 +10
10 + 5	20 +10	30 +10	40 +10

2   I   0   7

<u>6</u>   5   4   3

20   10   8   7

50   40   30   15